OUR UNIVERSE

OUR UNIVERSE

AN ASTRONOMER'S GUIDE

JO DUNKLEY

THE BELKNAP PRESS OF
HARVARD UNIVERSITY PRESS
CAMBRIDGE, MASSACHUSETTS
2019

Printed in the United States of America

Originally published in 2019 in the United Kingdom by Pelican

Book design by Matthew Young
Set in 11/16.15 pt FreightText Pro
Typeset by Jouve (UK), Milton Keynes
Books, an imprint of Penguin Books

First Harvard University Press edition, 2019

Library of Congress Cataloging-in-Publication data is available
from the Library of Congress

ISBN 978-0-674-98428-8 (cloth : alk. paper)

For my girls.

Contents

ACKNOWLEDGEMENTS

This book came about because of my friend and agent Rebecca Carter. She turned an idea into a reality, guiding and encouraging me to write. Chloe Currens and Tom Penn at Penguin, and Ian Malcolm at Harvard University Press, have been invaluable editors. Their many suggestions made the book much better, and I am particularly grateful to Chloe for steering me through to the finish. Thanks too to my US agent Emma Parry, and to the excellent Penguin production team.

I thank friends from university who would ask me questions about space and helped me discover the fun of explaining the wonders of the universe. I had them in mind when writing, especially Tom Harvey, Lou Oliver and Dan Smith. I also thank the students in schools I have visited, and people at public lectures, for asking such good questions. Many ideas for simplifying concepts in this book came from a teacher enrichment course in astronomy that I co-taught at Princeton in 2008 with science teacher Ilene Levine, with guidance from educator Lindsay Bartolone. My thanks to David Spergel for encouraging me to do it.

I am grateful to the Oxford physics department, where I was based until 2016, for making public engagement in science an integral part of our academic lives. Pedro Ferreira at Oxford showed me that it is possible to do research and write at the same time. Andrea Wulf introduced me to the wonderful story of the transit of Venus expeditions. I thank colleagues and fellow-astronomers for ideas or comments, including Neta Bahcall, George Efstathiou, Ryan Foley, Wendy Freedman, Patrick Kelly, Jim Peebles, Michael Strauss, Joe Taylor and Josh Winn. This book would not have been finished without the valued input of the Princeton Astrophysical Sciences graduate students. Goni Halevi, Brianna Lacy, Luke Bouma, Johnny Greco, Qiana Hunt, Louis Johnson, Christina Kreisch, Lachlan Lancaster and David Vartanyan all helped me check details and made suggestions that improved the book. Any remaining errors are mine.

Juggling research, writing and raising children has only been possible with the support of my husband, Fara Dabhoiwala, whose own writing achievements inspired me to try it out. He and my daughters and step-daughters make life more joyful; they are my universe.

INTRODUCTION

On a clear night the sky above us is strikingly beautiful, filled with stars and lit by the bright and changing Moon. The darker our vantage point, the more stars come into view, numbering from the tens or hundreds into the many thousands. We can pick out the familiar patterns of the constellations and watch them slowly move through the sky as the Earth spins around. The brightest lights we can see in the night sky are planets, changing their positions night by night against the backdrop of the stars. Most of the lights look white, but with our naked eyes we can notice the reddish tint of Mars, and the red glow of stars like Betelgeuse in the Orion constellation. On the clearest nights we can see the swathe of light of the Milky Way and, from the southern hemisphere, two shimmery smudges of the Magellanic Clouds.

Beyond its aesthetic appeal, the night sky has long been a source of wonder and mystery for humans around the world, inspiring questions about what and where the planets and stars are, and how we on Earth fit into the larger picture revealed by the sky above us. Finding out the answers to those questions is the science of astronomy, one of the very oldest sciences, which has been at the heart of philosophical inquiry since ancient Greece. Meaning 'law of the stars', astronomy is the study of everything that lies outside our Earth's atmosphere, and the quest to understand why those things behave the way they do.

Humans have been practising astronomy in some form for millennia, tracking patterns and changes in the night sky and attempting to make some sense of them. For most of human history astronomy has been limited to those objects visible to the naked eye: the Moon, the brightest planets of our Solar System, the nearby stars, and some transient objects like comets. In just the last 400 years humans have been able to use telescopes to look deeper into space, opening up our horizons to studying moons around other planets, stars far dimmer than the naked eye can see, and clouds of gas where stars are born. In the last century our horizon has moved outside our Milky Way galaxy, allowing the discovery and study of a multitude of galaxies that lie beyond our own. And just in the last few decades the technological advances of telescopes, and the cameras they use to capture images, have allowed astronomers to push our astronomical horizon yet further. We can now survey millions of galaxies, study phenomena such as exploding stars, collapsing black holes and colliding galaxies, and find entirely new planets around other stars. In doing so, modern astronomy continues to seek solutions to the age-old questions of how we came to be here on Earth, how we fit into our larger home, what will be the fate of Earth far in the future, and whether there are other planets that could be home to other forms of life.

The earliest known records of astronomy are more than 20,000 years old and take the form of carved bone sticks that track the phases of the Moon, used as ancient calendars in Africa and Europe. Archaeologists have found five-thousand-year-old cave paintings in countries including Ireland, France and India that record unusual events happening in the sky, including eclipses of the Moon and the Sun, and the sudden appearance of bright stars. There are also ancient monuments dating from that time, including Stonehenge in England, that may have been used as astronomical observatories to track the Sun and stars. The earliest written records of astronomy come from the Sumerians and later the Babylonians in Mesopotamia, in current-day Iraq. These include the very first catalogues of the stars, etched into clay tablets in the twelfth century BC. Astronomers were also active in ancient China and Greece by the first few centuries BC.

Though these first astronomers had only their eyes to use as tools, by the first few centuries BC the Babylonians had begun to identify the moving planets, distinguishing them from the fixed backdrop of stars and carefully charting their positions night after night. They began to keep systematic astronomical diaries, which led them to discover regular patterns in the movements of the planets and in the occurrence of particular events in the night sky, including the eclipses of the Moon. Nobody knew quite what those objects and events in the night sky were, but they could make mathematical models that could predict where the planets and the Moon would be seen, night after night.

Despite these considerable advances, great uncertainty remained as to how the heavenly bodies were configured and what they were made of. Which was at the centre of everything: the Earth, or the Sun? The realization that, in fact, neither was – that the universe does not have a centre – would only come many years later. In the fourth century BC the Greek philosopher Aristotle put forward a model, based on ideas by earlier Greek astronomers and philosophers including Plato, that put the Earth at the centre of the universe. The Sun, Moon, planets and stars were fixed in unchanging and rotating concentric spheres centred on the Earth. Aristotle presumed that the heavens were different to Earth in both composition and behaviour, imagining the celestial spheres to be made of a fifth, transparent element known as the 'aether'.

In the third century BC the Greek astronomer Aristarchus of Samos came up with the alternative suggestion that, in fact, the Sun might be at the centre of everything, and that it was light from the Sun that was illuminating the Moon. This heliocentric, or Sun-centred, model would better explain the observed motion of the planets and the changes in their brightness. Though we know now that this model is accurate, at least for our Solar System, Aristarchus' astronomical ideas were rejected during his lifetime, and would take over a thousand years to be accepted. Defenders of geocentrism, favouring an Earth-centred universe, had some apparently strong arguments on their side. For example, if the Earth moves, why do the stars not shift relative to each other as our viewpoint from a moving Earth changes? In fact, they do, but the movement is extremely slight because the stars are so far away. Aristarchus suspected this to be true but had no way to demonstrate it.

The erroneous Earth-centred model continued to prevail when it was adopted by Claudius Ptolemy, a highly regarded scholar from Alexandria in Roman Egypt, living in the second century AD. He wrote one of the earliest books on astronomy, the *Almagest*, which detailed forty-eight constellations of the known stars, along with tables that could predict both the past and future positions of the planets in the night sky. Many of these came from an earlier star catalogue of almost 1,000 stars compiled by the Greek astronomer Hipparchus. Ptolemy declared in his *Almagest* that the Earth must be at the centre of everything, and his influence was so great that the idea was to dominate for centuries. The *Almagest* was a central astronomical text for years afterwards and was expanded upon by generations of astronomers who followed.

During the Middle Ages, most progress in astronomy took place far from Europe and the Mediterranean, notably in Persia, China and India. In 964 the Persian astronomer Abd al-Rahman al-Sufi wrote the *Book of Fixed Stars*, a beautifully illustrated Arabic text detailing the stars constellation by constellation. It combined the star catalogue and constellations from Ptolemy's *Almagest* with traditional Arabic depictions of the imaginary objects or creatures traced out by the star patterns, and it includes the first report of our neighbouring galaxy Andromeda, at the time understood to be a smudge of light different in appearance from a regular star. During that same century, his compatriot, the astronomer Abu Sa'id al-Sijzi, proposed that the Earth rotates around its axis, thus taking a step away from Ptolemy's idea of a fixed Earth. Persia was home too of the great Maragheh observatory, a research centre founded by polymath Nasir al-Din Tusi in 1259 in the hills of Azerbaijan, which brought together home-grown astronomers with others from Syria,

Anatolia and China to make detailed observations of the movements of the planets and positions of the stars.

The sixteenth and seventeenth centuries brought a great revolution in astronomy. In 1543 the Polish astronomer Nicolaus Copernicus published his *De Revolutionibus Orbium Coelestium*, proposing that the Earth, as well as rotating on its axis, must also be travelling around the Sun, together with the other planets. His idea was strongly condemned by the Roman Catholic Church, which deemed the notion heretical; it would take sustained campaigning by a number of key figures, and new observations over a number of years, for it to be accepted eventually. The vital advance came with the invention of the telescope in the early 1600s.

Vision is enabled by light. The more light you can collect, the further out into space you can see. A telescope is, partly, a much larger bucket for collecting light than the human eye, allowing us to peer further out into the darkness of space and to see its features in better detail. It was Italian astronomer Galileo Galilei who first pointed a telescope at the sky in 1609, a primitive version that he had fashioned himself, which magnified the usual view of the sky by about twenty times. This was enough to let him see that Jupiter has its own moons, spots of light visible on either side of the planet that shift their positions as they orbit around. Without a telescope, or a pair of modern-day binoculars, they are hidden from view, too faint ever to discover.

In 1610 Galileo published his observations of Jupiter's moons, along with details of the Moon's uneven surface and his discovery of stars too faint to see with the naked eye, in his widely read *Starry Messenger* pamphlet. In it he supported Copernicus' view, boosted by the discovery of Jupiter's moons: they were a clear existence proof of celestial objects that did not orbit the Earth. Unfortunately, Galileo's evidence did not convince the Catholic Church: it remained strongly opposed to the Copernican description of the cosmos and would condemn Galileo, putting him under house arrest until his death.

Despite opposition from the Church, astronomers continued to make progress. German astronomer Johannes Kepler, who supported the ideas of Copernicus and Galileo, demonstrated in 1609 that the planets were all moving around the Sun following paths that have the shape of an ellipse: a squeezed circle. He also found that they followed a particular pattern that related their distance from the Sun with the time taken to orbit around it. The further away from the Sun, the longer it takes, but the distance and the time do not increase at the same rate: a planet twice as far from the Sun takes nearly three times longer to orbit. Later that century, in 1687, British physicist Isaac Newton would come up with his universal law of gravitation to explain why this pattern worked, in his famous *Principia*. His law stated that anything with mass attracts other things towards it, and the more massive the object, and the closer you are to it, the stronger the pull. If you are twice as close, you will feel four times the pull, and you will take less time to orbit. His law explained the patterns seen by Kepler, with the planets and the Sun orbiting around their shared centre of mass, and it showed that the laws of nature work the same in the heavens as they do on Earth. Observation and theory were now all in agreement, and an alternative to Ptolemy's celestial model was finally taken seriously across the world. The Earth really was moving around the Sun.

In the nineteenth century a second revolution in astronomy took place, driven by the invention of photography by Louis Daguerre in 1839. Before that, astronomical sketches had to be done by hand, which led to inevitable inaccuracies. And as well as being able to better measure the position and brightness of celestial objects, a camera can be set to a long exposure, allowing it to collect more light than an eye can see. In 1840 the English-American scientist John William Draper took the first photograph of the full Moon, and in 1850 the first picture of a star, Vega, was taken by William Bond and John Adams Whipple at the Harvard College Observatory. The 1850s also saw the invention of the spectroscope, a device used to split light seen through a telescope up into different wavelengths (which we will learn more about in chapter 2). These advances enabled astronomers to make extensive catalogues of stars in our Milky Way galaxy, including their positions, brightness and colours.

By the early twentieth century astronomers were building larger telescopes to see ever further into space. These were accompanied by key advances in our understanding of physics, including the development of general relativity by Albert Einstein and of quantum mechanics by Max Planck, Niels Bohr, Erwin Schrödinger, Werner Heisenberg and others. These new ideas allowed astronomers to make great advances in understanding the nature of objects in space, and the nature of space itself. Notable breakthroughs included Edwin Hubble's discovery in 1923 that our Milky Way is just one galaxy of many and Cecilia Payne-Gaposchkin's discovery in 1925 that stars are made primarily of hydrogen and helium gas (both of which we learn about in chapters 1 and 2).

Two technological advances of the twentieth century are of particular note, and both took place in the United States at the Bell Telephone Laboratories in New Jersey, a research and development company commonly known as Bell Labs. The first was the discovery in 1932 by Karl Jansky that we could observe radio waves coming from astronomical objects in space, opening up an entirely new window on the universe. This window was later expanded in the 1960s to include other types of non-visible light. The second major advance was the invention of the charge coupled device, known as a CCD, in 1969 by Willard Boyle and George Smith. Using an electrical circuit to turn light into an electrical signal, this device produces a digital image that we are familiar with from our digital phone cameras. They are more sensitive than photographic film, allowing astronomers to capture images of fainter and more distant objects in space.

Just in the past few decades there have been a wealth of advances in astronomical technology, theories and computation that bring us to our current state of knowledge. We have now seen all the way out to the edge of the observable universe, found millions of galaxies beyond our own, and have a coherent description of how our own Solar System, in our Milky Way galaxy, came to be here. The journey to our present-day understanding of the universe, and the many wonderful and strange things we now know about its workings, are the subject of this book.

*

As the scope of astronomy has grown, the nature of the astronomer has changed over the years. The title 'astronomer' is still the most generic, used for those of us who study and interpret what we see in the sky, but there are other titles too. Some of us call ourselves not 'astronomers' but 'physicists'. The usual distinction is that an astronomer studies the sky and makes observations of things in space. A physicist is a scientist interested in discovering the laws of nature that describe how things behave and interact, including the things in space. There is a great overlap between these two types of scientist, and no hard and fast way of defining the boundary. Many of us are both astronomer and physicist, and the name 'astrophysicist' is often used to describe someone working at the boundary of the two sciences. There are also different types of astronomers, depending on what questions are being asked. Some focus on the inner workings of stars, some on entire galaxies and how they have grown and evolved. The area of cosmology targets questions about the origins and evolution of the whole of space. One of the most rapidly growing branches of astronomy is that of exoplanets, the study of planets around stars other than our own.

Today, there are both professional and amateur astronomers. In the past, there was less of a divide between these groups. Ptolemy, Copernicus and Galileo all studied a variety of subjects. They and their successors followed such diverse pursuits as botany, zoology, geography, philosophy and literature, as well as astronomy. Now, the majority of new astronomical discoveries can be made only with professional-grade telescopes, too expensive for an individual to own and usually too large for an individual to operate.

To interpret in detail the phenomena we see through these telescopes can now take years of training. This means that we need professional astronomers, those of us who do little else in our working lives but study the universe. We are supported by universities, by governments, and increasingly too by philanthropists. Our demographic has also changed over the years, with more women in the field now than ever before.

Beyond the professionals, there is still an important role for amateurs to play. Small telescopes are still valuable for making particular observations, especially ones that need quick eyes on the sky to track unusual events that take place suddenly. There is also great demand for amateurs to help classify astronomical objects, using images taken by large telescopes and placed online. There is often too much data for the small professional community to process, and people are still better than computers at many tasks that require careful discernment of features, particularly unusual ones. In the past decade, amateur astronomers have found entirely new planets orbiting around other stars, and new and unexpected types of galaxies.

In broadening our horizons beyond our Solar System and the nearby stars, modern astronomy now has vast scope not only in space but also in time. We rely on light for access to space: we wait for light to arrive from distant places, and we see things in space because they either create light, or reflect it from another source. We then see them as they were when their light first set off. This adds another dimension to our observations of the sky: time. Light travels extraordinarily fast, 10 million times faster than a car on a motorway. This means that if you look at your nearest lamp, maybe a couple of metres away, you see its light a tiny fraction of a second in the past. The speed of light is almost irrelevant here. If instead you look at the Moon, about 200,000 miles away, you will see moonlight that is one second old by the time it reaches Earth. Light reaching us from the Sun is eight minutes old. Starlight is much older than that. The light from even our nearest stellar neighbour takes four years to reach us. When we look at the stars, we are looking back in time.

This is an incredible gift. We can see parts of space, parts of our universe, as they were many years ago. The further we can collect light from, the further back in time we can look. If you look at the bright star Betelgeuse, which glows in the Orion constellation, you wind time back more than six hundred years. Its reddish glow started its journey to Earth in the Middle Ages. The stars in Orion's belt are even further away. Their light, familiar to generations of humans, has travelled at least 1,000 years to reach us. This means we have a chance of understanding the history of the universe because we can see the more distant parts of it as they were in the past, thousands or millions or billions of years ago. This ability to look back in time has existed since humans first looked at the stars but has only become a key feature of astronomy in the past century as we have looked out beyond the Milky Way.

The great extent of the universe in both space and time can make modern-day astronomy seem overwhelming. Space is so immense that the numbers describing distances are at risk of becoming meaningless. Numbers with too many zeros are hard to process. To get around this, we come up with

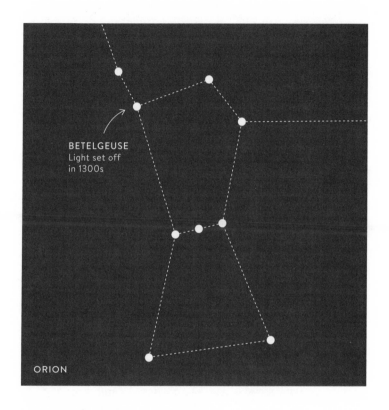

Figure 0.1
Stars in the Orion constellation; their light travels hundreds of years to reach us.

ways of making sense of the different scales of space, and we simplify things and let go of some of the details. We focus on knowing some of space, especially our own Solar System, very well and applying its lessons, where they seem relevant, to other areas. We content ourselves with knowing most of space less well. But some parts of distant space are particularly interesting and worth getting to know in greater detail, for example stars surrounded by planets that could resemble Earth, or galaxies where huge black holes collide or ancient stars explode.

This book is about our universe, which is the name we give to the whole of space that we know of, either that we can see with our telescopes or that we think is physically connected to the parts we can see. It will tell you what we think that universe is, and what it means to think about the whole of space and everything in it. It will give you a sense of how we on Earth fit into that bigger place. It will also tell you the broad-brush story of how our planet Earth came to be here, and what its future in our larger universe might hold.

We will not start the story at the beginning of the universe, though, because that is a rather unfamiliar place. Instead we'll start here and now, from our viewpoint here on Earth. In chapter 1 we put space in order. By looking deep into the night sky we know that things in space are not scattered at random. They have a definite pattern, a way things fit together from the smallest to the largest. We can step up from moons orbiting planets, to planets and asteroids orbiting stars, to collections of stars gathered into galaxies, and to the large clusterings of galaxies that are perhaps the largest objects in the universe. We will discover where Earth fits into that cosmic pattern and get a sense of the scale of space.

The second chapter tells the story of stars and how they live their lives. Some stars are just like our Sun, but many of them have very different life-stories. We will discover how stars make their light, and find the stellar nurseries where new stars are born. We will uncover the life and fate of our own Sun, and the more extreme lives of the largest stars that reach the end of their lives in violent explosions. Many of them end as dense black holes that will never allow light to escape. We will also find out about the extraordinary diversity of new worlds that are being discovered around foreign stars.

In chapter 3 we discover the wealth of invisible dark stuff in our universe that we can't actually see directly with our eyes or with our telescopes, even with telescopes that measure different sorts of light. This is a discovery that is less than a century old, and it has transformed our understanding of what the universe is and what it might be made up of. We are eagerly working to understand what it is, because it has a huge effect on all of the things that do shine brightly, and because at a fundamental level it appears to be part of the basic building blocks of nature.

In the fourth chapter we will find out how space has changed through the years. There are numerous galaxies beyond our own Milky Way, and they almost all appear to be moving away from us. We are led to the inevitable conclusion that space is growing, and that at some time in the past it likely had some sort of a beginning, the thing we call the Big Bang. We can now trace the evolution of the universe almost all the way back to that time and work out when that beginning happened. We will also encounter the idea of space itself having a shape, and the possibility of finding out if our universe is infinitely big.

The last chapter is a whistle-stop history of the universe. It takes us through the life of the universe from its first moments to where we are today. Tiny features imprinted at the beginning of the universe turn over billions of years into galaxies filled with stars, including our Milky Way, the home of our Solar System. Much of our understanding of what happened comes from combining observations with computer simulations that attempt to recreate how the universe could have evolved. Our own Sun and Earth formed when the universe was about two thirds of its current age; our Milky Way galaxy earlier still. We then look to what might happen next, both to our own part of the universe and to the whole of space.

We are living in an era with such unprecedented technological capabilities, both of our telescopes and our computers, that in our own lifetimes we can hope to take great leaps towards solving many of the outstanding mysteries in astronomy. We could find other planets that show signs of hosting life, discover what the invisible part of the universe is made of, and uncover how space itself began to grow. We might also find something completely unexpected, that could change the course of astronomy once again.

CHAPTER 1

Our Place in Space

Here on Earth we can define our location in a building, on a street, in a village or a town, in a country, on a continent, or in a hemisphere. We are, of course, part of something bigger too, and we can keep moving our horizons outwards to understand how we fit into this much larger place, our universe. In this chapter we will take steps ever further out until we reach the physical limits of what it is possible for us to see, and we will find out that Earth is just one of many places in the universe where life like us might have begun.

Our planet Earth spins around once a day, orbiting all the way around the Sun once every year. Its North Pole is tipped at an angle, such that the surface of the northern hemisphere receives the Sun's rays face-on, or closer to face-on, during daytime in the northern summer. For that part of the year, the Sun's rays are more densely concentrated on the Earth's surface in the northern hemisphere than in the south, and the north feels its most intense sunlight. Six months later, it is the southern hemisphere's turn to be angled towards the Sun, with the northern hemisphere tipped away and the North Pole now plunged into darkness.

In diameter, the Earth is almost 8,000 miles across, its size first worked out more than 2,000 years ago in ancient

Egypt by the Greek scholar Eratosthenes. It is curved like an orange, so the length of a person's shadow will depend on how far north or south they are. Eratosthenes noticed that the Sun was directly overhead in the town of Syene on the summer solstice at noon, so he went to Alexandria, a town known to be almost 600 miles due north of Syene, to measure the length of a shadow there at midday on that same day of the year. Knowing the distance between the two towns, and the height of the object casting the shadow, the length of the Alexandria shadow would depend only on the size of the Earth. A smaller planet would curve more between the two towns, making the shadow longer. This simple deduction led Eratosthenes to work out Earth's size to within about 10 per cent of the true value, a remarkable achievement at the time.

Earth's nearest neighbour in space is the Moon. Every month it orbits around us, close enough to pull our oceans gently towards it, swelling their tides usually twice a day as our Earth spins around. Our Moon is only a little over 200,000 miles away, far nearer than any of the planets in our Solar System. If we imagine our Earth shrunk to the size of a basketball, the Moon would be the size of an orange, travelling around it on a roughly circular path that would just fit inside a basketball court. The Moon happens to be just the right size and distance that when it passes right between us and the Sun, it can briefly block out all of the Sun's light in an awe-inspiring eclipse. This is a remarkable coincidence, as the Moon is 400 times smaller across than the Sun, but it is also 400 times closer to us, so both objects appear the same size in the sky. Eclipses are rare, though, because the

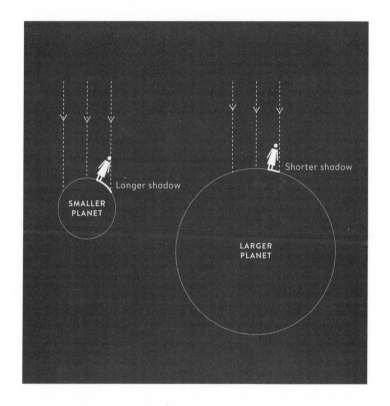

Figure 1.1
A person casts a longer shadow on a smaller, more curved, planet. Eratosthenes used this pattern to calculate the size of Earth, more than 2,000 years ago.

Moon's path around the Earth is not aligned with the Earth's path around the Sun. If it were, eclipses would happen every month.

The Moon is so close, of course, that humans have actually been there, although not for half a century. Only two dozen people have ever made the journey, on the legendary Apollo spacecraft, and only half of them set foot on the Moon's foreign surface. It would have been a dramatically different experience walking on the Moon than it is on Earth. The Moon is so small that the pull of gravity is six times weaker than here on Earth, so, unencumbered by a space suit, you could probably jump right over another person. Even wearing their bulky equipment, the Apollo astronauts are pictured in footage from the Moon landings bounding in leaps and hops across the lunar landscape.

Our Moon has a far side that is forever hidden from us on Earth. During its month-long orbit around Earth it spins only once, always keeping the same side facing us. This is quite unlike the Earth itself, which spins around every day of its year-long orbit around the Sun. Our Moon is almost as old as the Earth, born about 5 billion years ago. The most popular theory is that it formed from rocky debris left over after a violent collision between the newly formed Earth and another planet-sized object. Astronomers don't know for sure whether this happened, but if so, the Moon would have started out much closer to our Earth than it is now, appearing much larger in the sky. For its first few million years of life the young Moon would have spun around quickly as it travelled on its journey around Earth, showing first one side and then the next.

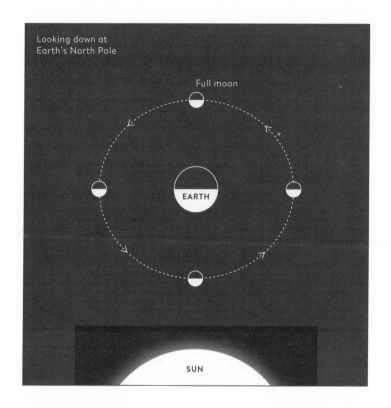

Figure 1.2

As the Moon orbits around Earth, half of it is lit up by the Sun. How much we can see of that illuminated half depends on the Moon's position.

The tides that the Moon creates as it orbits the Earth occur because the pull of Moon's gravity on the oceans nearest to it is stronger than its pull on the central core of our planet. This elevates the water level on the side closest to the Moon. The same thing happens to the oceans farthest away, on the opposite side of the Earth, since they are being pulled less strongly by Moon's gravity than the central part of Earth is being pulled. In most places the effect is two high tides every time the Earth spins around. And although the Moon does not have oceans, the Earth's pull of gravity has a similar effect on the Moon, slightly elongating it in the direction pointing towards the Earth. As the Moon spun around faster in its early years, the pull of Earth gently worked to keep the elongated part of the Moon directed towards it, slowing down its spin over millions of years until its familiar face was locked in place towards us, its far side no longer visible.

The Moon's pull of gravity has affected the spin of Earth in the same way too, gradually slowing it down by about fifteen seconds every million years. Days would only have been a few hours long early in Earth's life, and many years in the future the spin of the Earth may slow down so much that one of its sides permanently faces the Moon. And, since Earth spins on its own axis faster than the Moon orbits around us, the tidal bulges on Earth move slightly ahead of the Moon. This pulls the Moon around faster and is gradually moving it into a larger orbit: the Moon is moving a few centimetres away from us every year.

The Moon shines so brightly in the sky that it is easy to forget that it gives out no visible light of its own. It is, rather, illuminated by sunlight, but the Sun can only light up one side of the Moon at a time. That side is not always the same one that faces us on Earth. When the Moon is on the opposite side of the Earth from the Sun, it appears full, a complete bright circle. On the rest of its month-long journey around the Earth we can see only a portion of its sunlit side, and then briefly nothing at all.

We are so used to having one moon that it seems quite normal. But even in our Solar System it is not normal at all. Jupiter and Saturn each have more than sixty moons. Our outer neighbour Mars has two, but our inner neighbours Venus and Mercury have none. The presence of our single moon shapes life as we know it. Without the Moon we would have no tides, our days would be much shorter, and our regular seasons would most likely have been significantly disrupted. This is because it is the pull of the Moon that helps keep our Earth spinning at its fixed tilt compared to its orbit around the Sun. Far into the future, when the Moon has moved much farther away from us, our planet will likely wobble far more, tilting quite unpredictably as it journeys around the Sun.

Stepping outwards in space from our Earth–Moon pair, we come to the Solar System as a whole, the loosely defined collection of objects that are centred on our own star, the Sun. We know the Sun extremely well, of course, and the planets that orbit the Sun are also familiar to most people, at least by name. There are also asteroid rocks, comets, dwarf planets, and innumerable pieces of space debris that are drawn towards the Sun and orbit around it.

Despite all this, the Solar System is astonishingly empty. It can be hard to get a good sense of this, as pictures on the pages of a book do not easily capture the true scale of things. A convenient way to imagine the sizes is to shrink the Earth to the size of a small peppercorn, a couple of millimetres across. With the Earth this small, the Sun becomes a basketball, one hundred times larger from side to side. If we now put the basketball-Sun down and work out where Earth should be, we might expect it to be fairly nearby. But you would need to walk twenty-six large paces to reach the peppercorn-Earth, the full length of a tennis court. In between the real Earth and the Sun there are just two tiny planets, Venus and Mercury. Mercury would be ten paces from the Sun in this model, and peppercorn-sized Venus nineteen.

To reach our outer planetary neighbours starts to require some serious walking. Mars, a half-peppercorn-sized planet like Mercury, would be fourteen more paces past Earth. The largest planet, Jupiter, a large grape to Earth's peppercorn, would be almost 100 paces further. Jupiter is five times further from the Sun than Earth, or five tennis courts laid end to end. A little over another hundred paces on comes Saturn, an acorn, now ten times as far from the Sun as Earth. Uranus is twenty times and Neptune thirty times further than the Earth from the Sun. Tiny Neptune, like Uranus about the size of a raisin, is now almost half a mile away from our basketball-Sun, almost 800 paces and a roughly ten-minute walk. You could hold all those planets comfortably in your hand; the rest of the Solar System is almost entirely empty.

It is easy to imagine the planets all lined up neatly in a row: Mercury, Venus, Earth, Mars, Jupiter, Saturn, Uranus, Neptune; but of course they are not like that. At any given moment they are at different positions on their journeys

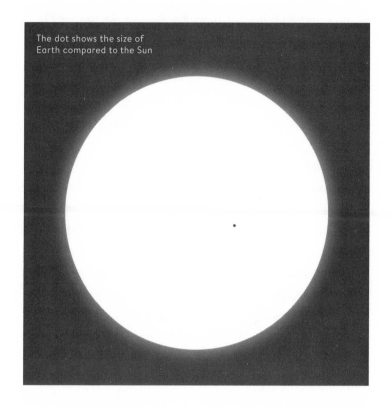

Figure 1.3
The size of Earth compared to the Sun.

around the Sun. They travel around at different speeds too, with longer years – or orbit times – the further they are from the Sun. A Mercury year is just three of our Earth months long. On Mars the year is almost twice ours; on Saturn it is almost thirty times longer. Birthdays would be a rarity in the outer reaches of the Solar System.

The position of the planets in the night sky is ever changing, as they each make their journey around the Sun. With our unaided eyes we can see five of them, Mercury, Venus, Mars, Jupiter and Saturn. Uranus and Neptune, further away, are simply too faint. Every night those five planets take a new position against the backdrop of stars. Sometimes they appear curiously close together, but this is just how we see them from Earth. Jupiter is many times further away from us than Mars, even when it appears right next to it in the night sky. Given the different speeds of the planets, the situation sometimes arises that just before dawn or just after sunset we can see both the inner and outer ones at the same time. Very rarely things align just right so that you see all five planets in the sky at once.

As well as tracking its movements through our sky, we can usually tell a planet from a star by its twinkle. On average planets twinkle much less than stars. That twinkle comes from the starlight or planet-light being jiggled around by variations in the air temperature in Earth's own atmosphere. The rays of light are bent, or refracted, by the air molecules on their journey to our eyes, and this makes the star look to us like it is continually moving around a tiny bit. We see that apparent motion as twinkle. The light from a planet does move around in the same way, but planets are so much closer to us than the stars that they appear larger in the night sky. Light coming from different parts of the planet's surface gets bent in different directions, and this reduces the overall twinkle.

We take for granted that we know the size of the Solar System, but it took years of effort and a rare alignment of the planets to measure the distance from Earth to the Sun, and from there the distances to all the planets. The earliest convincing measurements were made during the two transits of Venus in 1761 and 1769, when Venus passed right in between the Earth and the Sun. This is a wonderful story of intrepid adventuring and international scientific collaboration, and owes much to the impressive foresight of Edmund Halley, an astronomer at the University of Oxford, who is best remembered for his famous comet.

Transits of Venus are rare, occurring a little less than twice every century, since Venus and Earth do not orbit the Sun on the same flat plane. Most years Venus travels in between the Earth and the Sun, but does not cross our direct sight-line. It was the astronomer Johannes Kepler who first predicted the transits, calculating that both Mercury and Venus would cross the Sun in 1631. His predictions were confirmed, but Kepler died in 1630, never to see them for himself. In the following years British astronomer Jeremiah Horrocks worked out that the Venus transits should occur in pairs, separated by eight years. He was almost too late in discovering this, finishing his calculations only a month before the second transit was due to happen in 1639. Luckily he was just in time and, armed with his prediction, he successfully observed Venus crossing the Sun from his home in Lancashire.

In 1677 Edmund Halley travelled to the island of St Helena in the Atlantic to map the stars that were only visible from the southern hemisphere, and while there watched the transit of Mercury across the Sun. Inspired, he realized that a transit of Venus provided the key to finding out the size of the Solar System. His method uses parallax, which is an easy concept to understand as you can also use it to measure the length of your arm. You can do it by holding out your index finger at arm's length, closing one eye and noting where on the opposite wall or backdrop your finger appears to be. Then switch eyes, closing the other one instead. Your finger appears to move sideways. This movement is called parallax, and you can use it to work out how far your finger is from your eye, without actually measuring the length of your arm.

You should notice that if you hold your finger much closer to your eye, as if you have a shorter arm, your finger appears to move sideways a greater amount. The pattern is that the shorter your arm, the more your finger moves sideways. It is convenient to measure the amount of sideways motion as an angle that your finger has moved through. If you were to spin all the way around once, your finger would sweep through an angle of 360 degrees. You should find that your finger appears to move sideways by a few finger-widths when you switch eyes. For reference, a single finger or thumb held at arm's length takes up an angle of about 2 degrees from side to side.

If you know the distance between your two eyes, which might be a few centimetres, you can figure out the exact length of your arm. Here you are using triangle trigonometry. If you know the length of one side of a right-angled triangle, and the size of one angle, you can find out how long the other sides are. Here you have two right-angled triangles, back to back. They are each perhaps 4 centimetres long on their short sides (that is, between each eye and the bridge of your nose). If you measure the angle that your finger moves between closing one eye and the next, that is the same as twice the angle at the far tip of each right-angled triangle. So, if your finger moves 8 degrees, for example, you can work out that the length of your arm is almost 60 centimetres.

This has little practical use, of course, as there are easier ways to measure the length of your arm. But this same method can let us find the distance from Earth to Venus, and there it is invaluable. To use parallax to do this, your two eyes become two positions in the northern and southern hemispheres on the Earth, spread as widely apart as possible. Your finger becomes the planet Venus, which you are trying to find out the distance to. And the backdrop to your finger becomes the Sun. As with many other measurements in space, the triangles here are vast, their short sides being half the distance between the viewing positions on Earth.

You do the equivalent of closing one eye by looking at Venus from the northern hemisphere location as it transits the Sun and noting its position. Then you close the other eye by looking at Venus now from the southern hemisphere and once again seeing where it appears against the background of the Sun. Just like your arm, the more Venus moves against the backdrop of the Sun, the nearer it is to Earth. To work out the distance to Venus, you then just need to know the distance between your two observers on Earth.

There is a complication with this plan. The surface of the Sun is rather featureless, so back in the eighteenth century it

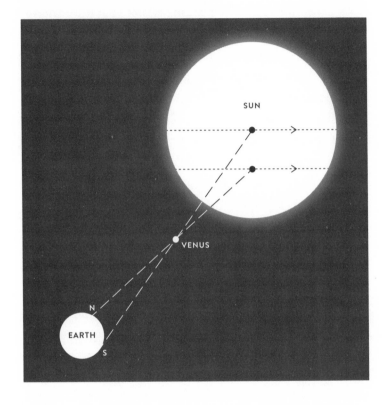

Figure 1.4
The time taken for Venus to transit across the Sun depends on where on Earth it is viewed from. The bigger the variation, the closer Venus must be to Earth.

would have been too difficult to accurately judge the precise position of Venus seen from different locations on Earth. Halley had an elegant solution, realizing that not only would Venus' position change depending on the viewing position, but also its transit time across the Sun. The two paths across the circular disc of the Sun would be different in length. The larger the difference in length, or the longer the difference in transit time, the larger the movement of the position of Venus would be against the Sun's backdrop, and the closer Venus would be to Earth.

This measurement would tell us the distance to Venus, and from there it would be a simple leap to get the distance to the Sun and to the other planets. Johannes Kepler had worked out the pattern that relates the orbit time of a planet with its distance from the Sun, with more distant planets taking longer to orbit. Astronomers had long known the length of a year on Venus by observing its changing position in the night sky. Knowing the length of a year both on Venus and on Earth, and the distance from Earth to Venus, was enough to set the scale of the whole Solar System.

Halley had worked out how to do it, but he knew that he would not live until 1761 to see the next transit of Venus himself. Undaunted, he left inspirational instructions to the next generation of astronomers, exhorting them to go and measure the transits:

> I recommend it therefore, again and again, to those
> curious astronomers who (when I am dead) will have
> an opportunity of observing these things, that they
> would remember this my admonition, and diligently

apply themselves with all their might to the making this observation; and I earnestly wish them all imaginable success. (Edmund Halley, 1716)

Halley's message worked. Almost twenty years after his death, astronomers from around the world came together to measure the transits. They were brought together by the French astronomer Joseph-Nicolas Delisle, who strongly encouraged the international scientific community to coordinate observations. Venus takes several hours to cross the Sun, and the difference in this transit time seen from widely spaced locations would be several minutes. To be able to make this measurement accurately, astronomers would not only need to visit far-flung locations in the two hemispheres of the Earth, but would also need to be able to accurately time the duration of the transit, measure their longitude and latitude accurately and have good enough weather to see it. This was not a project for a lone astronomer. Delisle managed to inspire hundreds of astronomers from the United Kingdom, France, Sweden, Germany, Russia and America to take part in this extraordinary coordinated endeavour, in what would be the first example of the global astronomy community in action.

In advance of the transit, Delisle had identified a drawback to Halley's method, which was that the entire transit had to be viewed. This was possible only on the section of Earth that faced the Sun for the full seven-hour duration and required good weather for that long. Delisle came up with an alternative method. Instead of viewing the entire transit from northern and southern locations, the start or end of the transit could be observed, as long as it was timed accurately and observed from various distant locations around the globe. Groups in the east would see Venus start its transit across the Sun before those in the west. A longer time-difference between various observers would imply a closer distance to Venus, equivalent to Venus moving further on the backdrop. Using Halley's method, the astronomers had to measure the duration of the transit accurately. Using Delisle's method, they simply had to record the start or end time accurately. Neither measurement was a small feat in those days and would rely on carefully transporting pendulum clocks as well as telescopes to distant locations.

In preparing for the 1761 transit, groups of astronomers took extraordinary journeys to attempt these measurements, travelling to South Africa, Madagascar, St Helena, Siberia, Newfoundland and India. Much has been written about their fascinating and challenging expeditions. Some travelled for months, only to have Venus obscured by clouds, and many got caught up in the Seven Years War, which had begun in 1756. The most accurate measurements in the southern hemisphere were made at the Cape of Good Hope by the British astronomers Charles Mason and Jeremiah Dixon, who had been sent by the Royal Society to reach Sumatra but diverted after their boat was attacked. Their success would lead them to their better-known jobs in the United States, surveying the border that would become known as the Mason–Dixon Line.

Many of the measurements made in 1761 were less accurate than hoped or ruined by bad weather. Once home, the astronomers combined the results together to estimate a distance to the Sun of between 77 and 99 million miles. It was lucky that they had the 1769 transit to refine their approach and have another chance at good weather, although Delisle himself did not live to see the second attempt. This time the British explorer James Cook was sent to Tahiti, and there were expeditions to Hudson Bay in Canada, Norway, Baja California, and Haiti. Many of these groups made successful measurements, and by combining all their timings the astronomers worked out the distance to the Sun to be almost 94 million miles, this time with an uncertainty of less than 2 million miles. They got it right: today's accurate measurement is 93 million miles.

Much of this story reflects how we still do astronomy today, more than 200 years later: working out how to make a difficult measurement, coming up with different ways to do it, planning many years ahead and going to often inhospitable and inaccessible places to get the best measurement possible. Central to the project's success was applying for funding from national governments for equipment, salaries and travel costs, coordinating with national and international teams, and combining results from different groups. These are all things we still do now in astronomy. Like today, each country's groups were happy to work together towards a common purpose but were particularly eager to make the best measurement themselves. As scientists we are often both competitive and collaborative in our pursuit of new discoveries.

The distance to the Sun, 93 million miles, starts to become a little awkward to handle given all the zeros. To simplify matters in measuring the vast distances of space, astronomers use much larger units of measurement. One of those is the Astronomical Unit, defined as the distance from the Earth to the Sun. The distance from the Sun to Neptune at the outer limit of the Solar System is, then, about 30 Astronomical Units, a much easier number to keep track of than 3 billion miles. We also measure distances in space by how long light takes to traverse them. In one hour light travels 700 million miles, so we call this distance a light-hour. The distance from the Sun to Neptune is then about four light-hours. This may be no easier to remember than 30 Astronomical Units, but this way of measuring distance is useful as it easily scales up to light-years, the distance that light can travel in a whole year. A year is made up of almost 9,000 hours, so a light-year is almost 9,000 times further than a light-hour, roughly 6 trillion miles. We don't need light-years to measure our Solar System, but as we move outwards in space, even to the nearest stars, it will prove useful to have such a huge unit of measurement to keep the numbers we deal with in check. Light-hours and light-years are also helpful in reminding us of the passage of time, as they indicate how long ago the light would have set off from the object we are looking at. Jupiter and Saturn are a few light-hours away from us, so our view of them comes from a few hours in the past.

With the scale of the Solar System in mind, we now focus on the tiny planets themselves. What are they actually like? First up is **Mercury**, closest to the Sun and orbiting in only three of our Earth-months. It is a very inhospitable place, without any atmosphere to protect it and in some places getting hotter than 400 degrees Celsius. The pull of the Sun's gravity has slowed down Mercury's spin, just as Earth has slowed down our own Moon. It now spins so slowly that a Mercury-dweller would spend one whole Mercury-year by day and then one year by night, spinning back to face the Sun only once every two years. During a night-time on Mercury, three Earth-months pass before sunrise comes around again. No wonder then that during that long period of darkness it gets extremely cold, reaching more than 100 degrees below zero. The chances of there actually being a Mercury-dweller to experience this radical climate are extremely small: it is very unlikely that there is any life there at all.

Mercury looks a bit like our Moon. It is covered in craters from old collisions with huge rocks that likely happened in the early days of our Solar System. Thirty years after their first mission to fly by Mercury in the 1970s, the National Aeronautics and Space Administration, known as NASA, sent the robotic Messenger spacecraft to Mercury in 2006. Entering into orbit in 2011, it took wonderful pictures of the planet's features and then ended its mission with a vigorous flourish in 2015 by crashing down into the surface. A new space mission called BepiColombo was launched in 2018 by the European and Japanese space agencies to study the planet in more detail, after a seven-year journey to get there.

Outwards from Mercury, our neighbour **Venus** comes next. Similar in size and weight to Earth, Venus also has its own atmosphere, albeit toxic by comparison with ours. It would be a very difficult place for any living creature to survive. Thick with carbon dioxide and clouds of sulphuric acid, its cloaking atmosphere makes it the hottest planet in the Solar System, hotter even than Mercury, with temperatures reaching more than 400 degrees Celsius. We cannot see through Venus' atmosphere with normal, visible light, so have to peer through it using cameras that can see with radio and microwave light, which has longer wavelengths than visible light. Inside we can see a planet with a rocky surface, like a desert, covered in mountains, valleys and highland plains. Venus also has many hundreds of volcanoes, more than any other planet. They do not appear to be erupting, but the European Space Agency's Venus Express spacecraft has detected continued activity in the form of lava flows.

Venus is also unusual in spinning what we would consider the wrong way. It travels around the Sun anticlockwise like the rest of the planets, but it spins around on its own axis clockwise if we look at it with its North Pole pointing upwards. This is the opposite direction to almost all the other planets in the Solar System and means that a Venus-dweller would see the Sun rise in the west and set in the east. We are pretty sure that Venus would have started out spinning the 'right' way round, but we do not know exactly what happened to send it the wrong way. Perhaps a large collision with another big rocky object could have changed its direction many years ago, or maybe even knocked it upside down. Similar to Mercury, it also spins much slower than Earth. A Venus-day is just over 100 Earth-days long, and about half the length of a Venus-year. A Venus-dweller, should one miraculously survive long enough, would regularly spend fifty Earth-days in continual darkness.

As well as the Venus Express, more than twenty spacecraft have visited Venus. Back in the 1970s and '80s a succession of Russian spacecraft visited Venus in the first ever landings on another planet. There have been no landings now for more than thirty years, but many spacecraft orbiting or flying past the planet have taken close-up looks. Most recently the Japanese space agency's Akatsuki mission entered orbit in 2015. Akatsuki very nearly missed its chance when it was launched back in 2010. It was supposed to enter orbit around Venus almost a year later, but its engines did not fire for long enough to put it in the right position. After waiting five long years in a holding pattern around the Sun, a quick rocket thrust set it successfully back on track, ready to tell us more about the extreme weather system on Venus. Venus' atmosphere rotates in just four Earth-days, very much faster than the planet spins, and the cause of this is unknown. Images from Akatsuki have now revealed unexpected features which may help explain this, including a jet stream wind blowing at nearly 200 miles per hour around the planet's equator.

Leaving Venus behind, we now cross over to the far side of the Earth from the Sun, to our best-known neighbour, **Mars**, setting for countless stories about alien civilizations. Half the size of Earth from side to side and just a tenth of the mass, the pull of gravity on the surface of Mars is three times weaker than ours. If you were on Mars you could take giant leaps three times higher than here on Earth. Mars spins around anticlockwise just like we do, and its days are about the same length as ours. It has two moons, but they are tiny compared to our Moon, only 7 and 13 miles across. Seen from Mars, the closer moon, Phobos, would look about a third the size of our Moon, but smaller Deimos is so far away that it would look little different than a star in the night sky.

Mars is ripe for exploring. It is rocky all over and covered in mountains, valleys and deserts. Its famous reddish colour comes from iron oxide on the surface, similar to the rust we find on metal. The atmosphere on Mars is rather thin, so we can easily peer down onto its surface. In the nineteenth century, astronomers thought they could detect the signature of water in Mars' atmosphere and infer its presence through features on the surface of the planet resembling canals or riverbeds. This fed speculation about the possibility that the planet may be home to intelligent life – the Martians who soon populated the pages of science fiction. The features were later proved to be optical illusions, but robot explorers that have touched down in recent decades have indeed found traces of liquid water, along with indications that Mars may have been covered with large oceans in the past. There is the tantalizing prospect that life could possibly exist there. Wonder at this possibility, together with Mars being near enough and solid enough to land on, have led to plenty of robotic missions by NASA's Mars Exploration Program. It would be an enormous challenge to send people to the planet, but in the coming decades that might well become a defining part of the international space programme.

Beyond Mars lies a swarm of smaller rocks, the **asteroid** belt. Small but plentiful, these are irregular lumps ranging from pebble-size to town-size with a couple of hundred much larger giants among them, circling all the way around the Sun. If you squashed all the rocks together they would make a single object smaller than our Moon. The biggest rock lurking in the belt is one of our Solar System's dwarf planets, Ceres, itself a sphere about 600 miles across. Ceres has been known for 200 years, discovered by Giuseppe Piazzi in Palermo in 1801. To start with, astronomers thought it was a planet, but it was small, so small that it looked like a star instead of a round circle when viewed through the relatively limited telescopes of the day. Soon afterwards astronomers discovered more of these small objects in the sky, with Pallas in 1802, then Juno in 1804, Vesta in 1807 and many more from the mid-1840s onwards.

Initially they were all referred to as planets, but in 1802 astronomer William Herschel had suggested that these new objects be called asteroids, meaning 'star-like' objects, instead. His suggestion to demote these new planets did not catch on right away, so these large rocky objects were still referred to as planets for a few decades. Over time astronomers and intellectuals, including the explorer and geographer Alexander von Humboldt, argued that the smaller objects ought to be dropped from the planet list, and in the 1860s they were classified as asteroids as Herschel had advised. History repeated itself just a decade ago when our current list of planets started growing again, as new objects, larger than asteroids, were discovered in the far reaches of the Solar System. Instead of including all these new objects as planets, astronomers elected to create the new category of dwarf planet that would also include Pluto.

Beyond the asteroid belt lies the giant of our Solar System, **Jupiter**. Unlike the rocky inner planets, Jupiter is a huge ball of gas, not a place you could stand on. Made mostly of hydrogen and helium gas, the planet is covered in swirling, multi-coloured gas layers tinged orange by gases like ammonia and sulphur, with its distinctive Great Red Spot, in fact a storm that has been going on for centuries, like a gigantic eye peering outwards. Eleven times larger across than Earth, it is 300 times heavier than us and spins around even faster, once every ten hours. Grand old Jupiter is shrinking as it gets older, getting steadily smaller by about an inch per year. It likely started out about twice as large across as it is now.

Jupiter likely played a defining role in the history of our Solar System. The ordering of our planets may once have been quite different. A popular scenario known as the Grand Tack has Jupiter starting out slightly closer to the Sun than it is now. It might then have gradually travelled in towards the Sun, to a position between where Earth and Mars are now. On its way it might well have destroyed rocky planets larger even than the Earth. The gravity of neighbouring Saturn could have then turned Jupiter back around, sending it out of the inner Solar System and back through the asteroid belt towards its current resting place, leaving behind large rocky fragments that eventually became Mercury, Venus, Earth and Mars. We do not yet know this for sure, but our new ability just in the past decade to study other solar systems around other stars is now teaching us more about what could have happened in our own. (We will get to that in the next chapter.)

No spacecraft can land on Jupiter, as it doesn't have a hard surface, but many have visited or flown past on their journeys to even more distant places. These include the Pioneer and Voyager craft in the 1970s and the Galileo spacecraft, which orbited Jupiter for a few years in the 1990s and even parachuted a small probe through its atmosphere, measuring winds of hundreds of miles per hour. Cassini sent

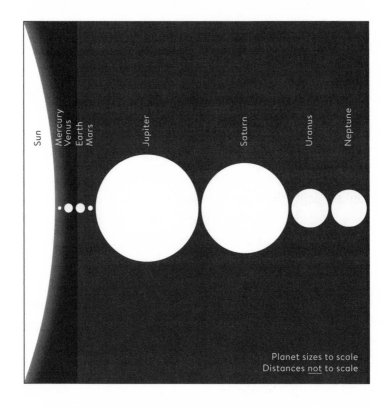

Figure 1.5
The relative sizes of the planets in our Solar System.

us beautiful images in 2000 en route to Saturn, and most recently NASA's Juno mission arrived at Jupiter in 2016, five years after setting off from Earth. As well as sending back stunning pictures from numerous fly-bys of the planet, including swirling cyclones surrounding its north and south poles, it is teaching us more about what Jupiter is made of and how it could have formed.

Jupiter has a wealth of moons. Its four largest, first seen by Galileo, can easily be seen with a pair of binoculars on a clear night, pinpricks of light lined up in a row either side of the larger disc of the planet. They are the perfect target for an astronomer starting to get to know our neighbours in the night sky. Jupiter's largest moon, Ganymede, is even bigger than Mercury. Europa, second-closest to the planet of the largest four, is particularly fascinating as it is enveloped in an ice-covered ocean that is likely liquid and is one of the most promising places for other life to have developed in our Solar System. This enticing possibility – that Jupiter's moons could harbour tiny life forms – is inspiring future missions to explore them in more detail. The European Space Agency is due to launch the Jupiter Icy Moon Explorer spacecraft in the early 2020s to study Ganymede, Callisto and Europa. NASA's Europa Clipper mission, due to launch on a similar timescale, will survey Europa's icy shell and help select a landing site for a future mission.

Stepping onward from Jupiter, we come to **Saturn**, the other marvellous giant of our Solar System. The rings that surround it mystified their first observers – Galileo described them as looking like ears or arms and speculated that they were moons – and make it perhaps the most distinctive of all our planets. The Cassini mission has shown us stunning images of the planet but it is also still a thrill, even for seasoned astronomers, to see Saturn and its rings by looking directly through a telescope eyepiece.

Like Jupiter, Saturn is a giant made up mostly of hydrogen and helium. There is probably a core of rock at its centre, then layers of liquid hydrogen, then an outer gas layer, with ammonia giving the planet its distinctive yellowish colour. Saturn is the only planet in the Solar System that on average is less dense than water, so would float in a large enough body of water. From afar the planet's rings look solid, but up close they are revealed as a thin disc that ranges from only about 10 metres thick to about a kilometre in some places, made up of ice, rocks and rock dust. The rings themselves extend as far as 50,000 miles from Saturn's equator. The origin of the rings is not certain, but they might have been left over from a destroyed moon or pieces of comets and asteroids.

Saturn spins every ten hours and, being ten times as far from the Sun as Earth, is a cold place, with winds reaching over 1,000 miles per hour. Like Jupiter, the planet has more than sixty moons, and they are similarly fascinating. The two that have attracted most attention are Titan and Enceladus for their relative similarity to Earth. Titan is Saturn's largest moon, slightly larger than Mercury, and it has its own nitrogen atmosphere as well as lakes and seas of hydrocarbons, islands, mountains, winds and rain of liquid methane. It is much colder than Earth, with an average surface temperature close to −200 Celsius. Enceladus is similarly cold but likely more hospitable: fly-by views have showed it to be possibly the most habitable place in the Solar System after

Earth, with only a thin icy crust covering the sea in places, and water plumes shooting out from the ocean deeps.

Out beyond Saturn lie elusive **Uranus** and then **Neptune**, twenty and then thirty times as far from the Sun as the Earth. They are the least explored planets of our Solar System. They both have atmospheres with a lot of hydrogen and helium, but also water, ammonia and methane. Uranus is layered in clouds blown by strong winds, and underneath those clouds is likely a fluid layer wrapped around a rocky core. An oddity of Uranus is that it lies almost completely sideways, its north and south poles lying in the same plane as its orbit. This orientation might have been established when the Solar System was forming, likely caused by collisions with other planets, and results in each pole getting around forty Earth-years of continual sunlight, followed by forty of continual darkness.

Uranus was not well known to ancient observers of the night skies. It is so dim and moves so slowly that it had been mistaken for a star. It was British astronomer William Herschel who, in 1781, discovered Uranus and initially classified it as a comet. He had not immediately been sure what it was, but he knew it changed its position in the night sky so could not be a star. Further calculations by Johann Bode and Anders Lexell soon showed it to be moving like a planet in orbit around the Sun. This new planet, the first to be found with a telescope, started out with an identity crisis, as Herschel initially named it the 'Georgian planet', in honour of the British king of the time, but unsurprisingly this choice was not very popular outside Britain. The name Uranus was finally fixed on in the 1850s.

Astronomers soon realized that, although Uranus clearly orbited the Sun, its orbit was odd, as if the pull of gravity from an unseen object was tugging on it. In the 1820s French astronomer Alexis Bouvard suggested that the pull might come from another celestial body, yet undetected. The astronomers Urbain le Verrier in France and John Couch Adams in the UK each figured out where it should be in 1846, having noticed that there were differences between Uranus' observed orbit and the prediction from Newton's laws of gravity. Le Verrier sent his prediction of the expected position of this new body by letter to the astronomer Johann Galle in Berlin. On receiving the letter, he found Neptune at nearly exactly the expected position that same night, a beautiful example of scientific prediction and observation working together. In many ways Neptune is similar to Uranus but has more visible weather patterns and is positioned the 'right' way up. Surrounded by clouds of ammonia and an atmosphere of hydrogen and helium gas, with some methane, which gives it its bluish colour, Neptune is mostly made of liquid, with a rocky core about the size of the Earth.

Out beyond Neptune lives the **Kuiper Belt**, the second main asteroid belt in our Solar System, and the dwarf planet **Pluto**, along with at least a few more dwarf planets. Pluto is mostly ice and rock, recently featured in close-up pictures taken in 2015 by NASA's New Horizons mission. The story of Pluto's discovery is instructive as to the importance of serendipity in science. In the late 1800s astronomers had studied the orbits of Uranus and Neptune and speculated that yet another planet might exist that was disturbing their paths. So in the early 1900s the search started for a ninth planet, nicknamed Planet X, at the Lowell Observatory in Arizona. Astronomer Vesto Slipher suggested to his young colleague Clyde Tombaugh that he look for objects that changed their position using photographic plates taken of the sky at different times. Tombaugh searched for a year and in 1930 found a new object that appeared to be moving. It was declared to be a new planet and named Pluto, after the Roman god of the underworld, following a suggestion by the English schoolgirl Venetia Burney to her grandfather, who had been head of the Bodleian Library at the University of Oxford. Her proposed name was relayed to America by an astronomer at Oxford; the eleven-year-old would thus become the only woman in the world to have named a planet. But, even though Pluto was real, it turned out to be a coincidence that it was found: it was not in fact big enough to have affected Neptune's orbit, and refined calculations showed that there was no need for a missing Planet X after all.

Pluto has been a source of great debate in recent years. Astronomers elected to demote it from planet to dwarf planet in 2006, voting together at a meeting of the International Astronomical Union in Prague. This was a difficult choice to make, since Pluto was then well known as one of the Solar System's nine planets, learned about by school children everywhere. But Pluto differs from the rest of our planets in important ways, in particular because it is so small. Astronomers came up with a formal definition that a planet should be a round object orbiting the Sun, large enough such that no other objects of similar size, apart from its moons, lie in the same orbit. All our other eight planets fulfil these requirements, but Pluto does not.

Pluto also has at least four similar dwarf-planet siblings, with Ceres known in the asteroid belt from the 1800s, and Haumea, Makemake and Eris found in the outer Solar System in 2004 and 2005 by a team of astronomers led by Mike Brown at the California Institute of Technology. There are likely many more out there, with hundreds of candidates already discovered. This decision to change Pluto's status and create the new category of dwarf planet mirrors the decision made by Herschel and others in the 1800s when the asteroids and planets were distinguished. As our scientific understanding grows, we need to be able to adapt the way we classify astronomical objects.

Despite Pluto's demotion from the planetary ranks, there may yet be nine planets. The history of Neptune's discovery may be repeating itself today. In 2016, Mike Brown and his colleague Konstantin Batygin predicted the existence of a new planet after close examination of the orbits of the small dwarf planets and asteroids beyond Neptune. Appropriately dubbed 'Planet 9', it is expected to be ten times heavier than Earth, similar in size to Uranus and Neptune, and much further out than our other planets. If the calculations are right and it does exist, it would take this planet more than 10,000 years to orbit the Sun, and for most of that time it would be twenty times further from the Sun than Neptune. Not all astronomers are convinced by the argument that Planet 9 should exist, but the search for it is underway.

Reaching the edge of our Solar System, we now journey much further outwards, to the stars seen in the night sky that inspire so much of our wonder about space. They are beautiful and mysterious, at once familiar and unknown. Aside from the Moon and the planets, they are also the objects in space that we can see most easily with our own eyes from our own back gardens. The number of stars that we can see, though, depends enormously on the light pollution created by human activity. In the city, we see only a thin scattering of the brightest stars. Looking at the night sky from the dark countryside shows the real richness that lies above.

It was the Greek astronomers Hipparchus and Ptolemy who, 2,000 years ago, introduced the ranking that we still use today for describing the brightness of stars. Stars appear to have slightly different sizes in the night sky, with the brighter stars appearing larger. The largest and brightest were thus termed 'first-magnitude' stars by the Greek astronomers. Stars ranked with increasing magnitude were smaller and fainter, up to the faintest that humans could see by eye on a dark night: sixth magnitude. By the eighteenth century astronomers realized that the stars are all far enough away to appear point-like; the brighter ones simply appear larger when viewed with our eyes or through telescopes. The name 'magnitude' stuck, however, and in 1856 the British astronomer Norman Pogson formalized the scale. He determined that the sixth-magnitude stars were 100 times fainter than those of first magnitude, so he proposed a scale such that a star, or other astronomical object, is a single step higher in magnitude if it is two and a half times fainter; an object five steps up in magnitude is then a hundred times fainter (or, two and a half multiplied by itself five times). Pogson's scale is still in use in modern astronomy.

Many of the stars that we can see at night are part of our

'Solar Neighbourhood', a name for the collection of stars that live nearest to our Sun and that travel around together within the larger Milky Way galaxy. To reach these stars we have to imagine leaving our Solar System and taking a big step out. Our nearest stellar neighbour is Proxima Centauri, part of a system of three stars named Alpha Centauri, and it is just over four light-years away from us. This means that the light we see now from that star set off four years ago and has only just reached Earth. That also means, curiously, that if anything had happened to Proxima Centauri in the past four years, we would have absolutely no idea about it, not yet. It also means that if we were to send a spacecraft like Cassini to the star it would take more than 50,000 years to get there, assuming a journey that long could somehow be fuelled.

The other stars we can see with our eyes are between a few and a few thousand light-years away from us, and number in the several thousands. Our Solar Neighbourhood includes just those that are tens of light-years away, numbering about a hundred stars. Remember that Neptune is four light-hours away from the Sun, so these star distances are many times larger. For comparison, if we were to scale down the whole Solar Neighbourhood to fit into our basketball court as we did before for the Solar System, we would now find that the entire Solar System would be as small as a grain of salt.

How do we know how far away the stars are if we can never reach them? Before the 1600s, astronomers found it improbable that there would be such a large gap between Saturn, which was the most distant known planet at the time, and the 'eighth sphere' of the stars as imagined by Copernicus. Progress came from the Dutch physicist Christiaan Huygens, an impressive polymath who, among his many contributions, designed a telescope to study the rings of Saturn, discovered Saturn's moon Titan, developed a theory of light and would also invent the pendulum clock. In 1698 he used the brightness of stars to attempt to work out how far away they are. Assuming all stars to be equally luminous if one were standing right next to them, he could compare the perceived brightness of Sirius, the brightest star in the sky, to our Sun. The further away, the dimmer it would appear. He came up with an ingenious scheme to work out how many times fainter he needed to make the Sun's light to match the brightness of Sirius, by obscuring the Sun with a piece of brass drilled with a tiny pinhole covered with glass. He estimated Sirius to be almost a billion times fainter than the Sun, which would put it 30,000 times further away from us than our Sun, or half a light-year away. The true distance is almost nine light-years. Huygens' work was impressive, but he got it wrong because he did not know that Sirius is intrinsically many times brighter than our Sun.

A more accurate method uses parallax, like we used for measuring the distance from Earth to Venus. To use parallax for stars, your two eyes now become the Earth at two times in the year, spread six months apart, when it is on opposite sides of the Sun. Your finger becomes the nearby star that you are trying to find out the distance to. And the backdrop to your finger becomes the backdrop of even more distant stars, which are so distant that they do not appear to move when the Earth moves around the Sun.

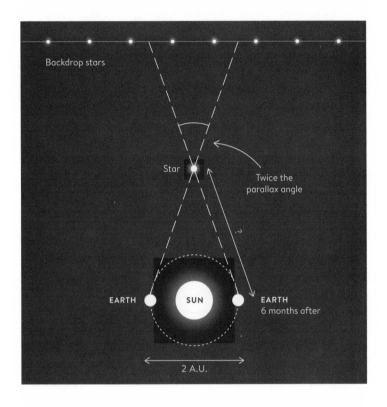

Backdrop stars

Star

Twice the parallax angle

EARTH

SUN

EARTH
6 months after

2 A.U.

Figure 1.6
Using parallax to measure the distance to a star. As Earth orbits
around the Sun, the star appears to move sideways. The more it
moves, the closer it must be to Earth.

For the distance to Venus we did the equivalent of closing one eye by looking at Venus transiting the Sun from one side of the Earth. Now, for stars, you close one eye by looking at the nearby star from your chosen observing spot on Earth and measuring where on the distant-star backdrop it appears. Then six months later you close the other eye by looking at the star again, still from Earth but now at the other side of its orbit, and once again seeing where it appears against the background. Just like your finger against the distant wall, the more the star moves against the backdrop, the nearer it is. To work out the distance to the star, you then just need to know the distance between your two Earth-eyes, which we already found out is two Astronomical Units, or a bit less than 200 million miles.

This idea is simple, but it is hard to put into practice. A star four light-years away from us moves only a tiny amount against the background stars. The angle is a little less than an arcsecond, which is an extremely small unit of measurement. (There are 60 arcseconds in an arcminute, and 60 arcminutes in a degree. To get a sense of how tiny an arcsecond is, remember that your thumb held at arm's length takes up about 2 degrees from side to side, which is a little over 7,000 arcseconds.) To see a shift that small requires an excellent telescope. Astronomers tried to use parallax in the 1700s, but it was just too challenging. But by the early 1800s technology was progressing, and in 1838 the German astronomer and polymath Friedrich Bessel successfully measured the distance to 61 Cygni, a binary system made up of two stars orbiting around each other. He discovered that it was 10 light-years away, a measurement that was followed later that same year by Friedrich Struve's measurement of 25 light-years to the star Vega, and Thomas Henderson's four light-years to Alpha Centauri.

This way of working out distances has produced astronomers favoured unit of distance, a parsec. This is the distance to an object that has a parallax of one arcsecond, which means it moves sideways by an angle of twice that, 2 arcseconds, when the Earth moves to the opposite side of the Sun. It is the equivalent of just over three light-years. Using telescopes on Earth we can only measure the distance using parallax to objects out to about 100 parsecs, or about 300 light-years, which lets us reach the stars in our Solar Neighbourhood but not much beyond. If we leave our atmosphere though, we can do much better. The Hubble Space Telescope can reach distances of up to 10,000 light-years using parallax, and since 2013 the European Space Agency's Gaia satellite has been measuring parallax distances to stars up to many tens of thousands of light-years away. In 2018 the Gaia team published a stunning catalogue with the distances to more than a billion stars.

Our Solar Neighbourhood is one small corner of a much larger place that we can call home. Many of the stars that are most familiar to us, like Alpha Centauri, Sirius and Procyon, inhabit the neighbourhood. But many others that we know best from the night sky, like the North Star and the stars in Orion's belt and the Little Dipper, are far beyond, lying hundreds of light-years away from us. They are part of our larger home, our Galaxy (which takes a capital 'G' to distinguish it from other galaxies and is also known as the Milky Way), an enormous collection of around 100 billion stars held together by the pull of their collective gravity. It is vast and magnificent, a huge spiral disc gently spinning around.

If we could look down on the starry disc of our Galaxy from above, we would see four star-filled arms spiralling clockwise inwards towards a brighter core of light at the middle, a little like water swirling around a plughole. Taking much longer than the planets to orbit in our Solar System, our Galaxy spins around every couple of hundred million years. Our Solar Neighbourhood sits in one of the spiral arms, known as the Orion arm, about halfway out from the middle of the Milky Way to the edge of the disc of stars. As the Galaxy spins around we move along with it a bit like a horse on a carousel, our whole Solar System travelling at an astonishing 500,000 miles per hour.

Our Galaxy is a truly vast place. It would take light about 100,000 years to cross from one side of the disc of stars to the other, and about 1,000 years from top to bottom. Remember that it takes light 'only' a few tens of years to cross the Solar Neighbourhood, and a few hours to cross the Solar System. If we were now to try to fit our whole Galaxy into the same basketball court that previously housed just the Solar Neighbourhood, we would find the Solar Neighbourhood shrunk to the size of a peppercorn. It would be moving clockwise round the room, roughly halfway out from the centre.

Of course, what is clockwise or anticlockwise depends on how you define up and down. As astronomers we have settled on a direction that we call 'up' for our Galaxy, and it is on the side of the disc that is aligned nearest to the 'up' of our Solar System, in the direction north of the North Pole of Earth. But those two 'up' directions are not the same, as the flat plane that is home to the Solar System's planets is not aligned with the flat disc of the Galaxy. If you point the first finger of your hand away from you with your thumb comfortably outstretched upwards, the Galaxy-disc would be lined up with your finger, and the Solar-System-disc would roughly line up with your thumb.

We will never be able to see the Milky Way from above. It is simply too big for us, or any craft we could invent, ever to leave it. We can look at it from within, though, and work out what it might look like to some being looking down at us from afar. On a clear night, the starry disc of the Galaxy appears as a strip of hazy light in the sky, this path in the sky inspiring our Galaxy's evocative name. The light is fainter than the brightest stars in the sky, so we can rarely see it in urban settings.

Why does it look like this? We can imagine the disc of stars a little like a saucepan lid, with our own Sun embedded in it about halfway out from the central lid handle to the edge. What do we see when we look in the sky? We imagine ourselves in the lid, and imagine the stars filling the saucepan lid. We see the nearby stars scattered in all directions around us, all over the sky, because our nearby stars surround us from all sides. And when we look in the directions that run directly through the saucepan lid, through the Galaxy's disc of stars, we are seeing light from a huge number of stars, which show up as a glowing strip of brighter light. When we look out in any other direction from the saucepan lid we are looking through a far smaller number of stars, and we see only darkness beyond the bright individual stars. The most distant star we can see as an individual object with our eyes is a few

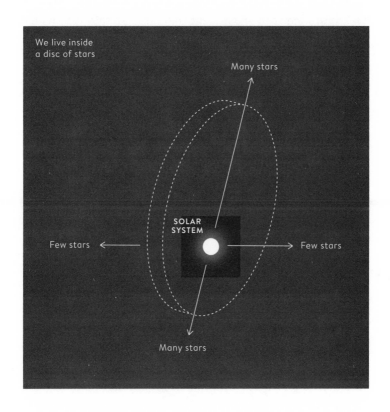

Figure 1.7
Why we see the Milky Way as a band of light in the night sky.

thousand light-years away. As it is nearly 30,000 light-years from us to the centre of the Milky Way, the glow from the collection of more distant stars can only be resolved into actual stars using telescopes. It was Galileo, back in 1609, who first identified stars in the band of light.

The Milky Way, of course, contains much more than just stars. We now think that at least half of the stars have their own planets. Our Galaxy is teeming with other stuff too, including tiny grains of cosmic dust, clouds of hydrogen and helium gas, black holes and other weird and wonderful things. We will get to those in the next chapters. The cosmic dust has an important impact on what we can see, as much of it lies in the way of the starlight, absorbing the light and darkening what would otherwise be a much brighter band of starlight from the Milky Way. Without it the night sky, particularly the light from the Milky Way, would be much more luminous.

We do not know quite how old our Galaxy is. It probably first formed when the universe was very young, more than 13 billion years ago. We know that because there are stars in our Galaxy that are almost as old as the age of the universe. The Galaxy would not have looked the same back then, though, as the disc of stars likely took a long time to emerge, and it would have collided and mixed with other galaxies in its past. It probably took on its now familiar form about 10 billion years ago. We will return to this idea in chapter 5.

It has taken novel ideas to discover exactly how large our Galaxy is. We have already seen that parallax, which relies on measuring how much stars move against their backdrop when the Earth moves round the Sun, takes us only about 300 light-years out from Earth, although much further using space telescopes. Here we are talking about seeing out to the backdrop stars themselves, up to 100,000 light-years away. How do we know how far away they are? They do not appear to move at all on the night sky as the Earth travels around the Sun. Some other method would have to be devised.

Henrietta Swan Leavitt, an astronomer at Harvard in the early twentieth century, was a key figure in working out a new approach. She was studying a type of star known as a Cepheid, which continually pulsates in brightness, appearing dimmer and brighter as it pulsates in size. In 1908 she discovered a pattern in these stars, figuring out that the brighter the star was at its source, the longer it took to pulsate. A very bright one would pulsate every few weeks or months. A faint one could take just days.

This was a game-changing discovery, made even more remarkable by Leavitt's own story. Born in 1868, she studied at Oberlin College in Ohio and then at what was later known as Radcliffe College. This was the women's college attached to Harvard (which did not admit women at the time). She studied music and then discovered her interest in astronomy, but after graduating she became ill and was left severely deaf. Her passion for astronomy continued, and she started volunteering at Harvard College Observatory in 1895, working for the astronomer Edward Pickering. A few years later she was hired onto the staff as one of many women 'computers' who were paid about thirty cents an hour to study photographic plates of light from stars. One of the goals of Pickering's programme was to catalogue the brightness of all the known stars. As a woman, Leavitt was not allowed to operate a telescope herself, nor was she given the intellectual freedom to explore her own theoretical ideas. Instead she was tasked with looking for stars whose brightness varied, by painstakingly comparing photographs taken on different nights. She was so good at her work that, among her many achievements, she discovered more than 2,000 variable stars in her lifetime.

Leavitt found the pattern in the Cepheid stars' pulsations, now increasingly known in the astronomical community as Leavitt's Law, using stars in the Magellanic Clouds. These are two objects that appear to the eye in the night sky as white smudges that you might indeed easily mistake for clouds. Now known to be galaxies separate from ours, at the time they were thought to be a collection of stars within the Milky Way. The Magellanic Clouds are visible only from the southern hemisphere, and Leavitt used photographs of them taken in Peru to find twenty-five Cepheid stars that all showed this particular relationship between the brightness and the regularity of the pulsations. The pattern showed up more obviously in these stars because they were all about the same distance from Earth.

Leavitt's discovery was extremely useful. Even if we cannot physically get to a star, we can look at it with our telescopes, and time how long it takes to pulse. Knowing the relationship between pulse-time and brightness, we can then deduce how bright the Cepheid star would be if we happened to be standing right next to it. This makes it something we call a standard candle in astronomy, something with a known intrinsic brightness. Using our telescopes to measure how bright the star appears to us here on Earth, we can work out how far away it is. The further away it is, the dimmer the star will seem.

Working with these Cepheid stars takes us out to the furthest reaches of the Milky Way and beyond and is still one of the most important ways of measuring distances in our universe. It was American astronomer Harlow Shapley, head of the Harvard College Observatory from 1921 to 1952, who used the pulsating stars' behaviour in the years leading up to 1919 to measure the size of the Milky Way and to find the position of our Solar System within it. Using the 60-inch telescope at Mount Wilson in California, then the largest telescope in the world, he found the distances to stars grouped together in 'globular clusters', dense collections of hundreds or thousands of stars held together by their gravity, spread throughout the Milky Way. For this he used a class of stars that pulsate even faster than the Cepheids, known as RR Lyrae stars, which have a similar pattern relating pulsation rate and brightness. They revealed the stars in the Milky Way to be a flattened disc over 100,000 light-years from side to side, with the Solar System about halfway out from the centre.

Until about a century ago people thought that our Galaxy might make up the whole of the universe. We now know that our Galaxy is not alone at all. It is one of very many, perhaps infinitely many galaxies. Together they group themselves into communities, the cosmic equivalent of towns and cities. Some galaxies live in the universe's small towns, collections of up to around a hundred galaxies known as galaxy groups. Other galaxies live in larger groupings of hundreds to thousands, known as galaxy clusters. Our own Milky Way is a small-town galaxy, living inside a group with more than fifty other neighbouring galaxies, which we call the Local Group. It is gravity that draws the members of our group together, the attractive pull of all the mass in all the stars and other stuff inside those galaxies.

We now know that two of our nearest neighbours are the Magellanic Clouds. The galaxies are far enough below the Milky Way that only people in the southern hemisphere are lucky enough to see them. For those of us who live in the north it is quite a thrill to take a trip to the southern side of our planet and suddenly spot them in the sky. They got their name in the 1500s from their first sighting by Europeans during Ferdinand Magellan's epic voyage from Spain to find a new route to the Spice Islands. Of course, they were well known to those living in the southern hemisphere long before then, appearing in oral traditions of indigenous groups in Australia, New Zealand and Polynesia.

Our whole Local Group is about 10 million light-years across, which is about a hundred times further across than the Milky Way. Aside from our own, there is only one large galaxy in the Local Group, another spiral galaxy called Andromeda. It sits over 2 million light-years away from us, more than ten times as far away as the smaller Magellanic Cloud. At about 200,000 light-years across, Andromeda is about twice our size and contains as many as a trillion stars. If we now imagine fitting our Local Group into that same basketball court that we keep returning to, our Milky Way would be about the size of a CD, and Andromeda a large saucepan-lid, and they would be about 3 metres apart. On this scale, the Large Magellanic Cloud would be about the size of a grape, the Small Magellanic Cloud a peanut.

Andromeda is one of the most distant objects we can see in space with our own eyes. Without a telescope only the brightest part at the middle is visible, so it looks a lot like a star. With a good telescope you can see its whole disc, appearing larger than the Moon and roughly the width of your three middle fingers held at arm's length. You can find it in the night sky in between the Pegasus and Cassiopeia constellations. The stars in those constellation patterns are all in our Milky Way, but they can guide our eyes to the more distant Andromeda.

Andromeda has been a known feature of the night sky since at least the tenth century, appearing in the *Persian Book of Fixed Stars*. In it Andromeda is first referred to as a 'small cloud'. Andromeda was later identified by telescope in the 1600s by the German astronomer Simon Marius, a contemporary of Galileo and responsible for naming Jupiter's four large moons. Andromeda then appeared in an important catalogue by the French astronomer Charles Messier, who began his career working for Joseph Delisle, the French naval astronomer who had played such an important role in the transit of Venus measurements. Messier's book, published in 1760, catalogues just over 100 objects that he referred to as nebulae, diffuse objects in the sky that clearly were not individual stars. What they were remained a mystery. Even at that time, no one would have known that Andromeda or any other of these objects were actually outside our own Galaxy.

Andromeda and the Milky Way are more than 2 million

light-years apart at present, but astronomers are fairly certain that in a few billion years the galaxies will collide. We are flying towards each other at a speed of more than 50 miles per second, and there is no turning back. Galactic collisions are major events, but they do not necessarily create havoc for the planets and stars that make up the galaxies. In the context of outer space, the stars themselves are so small, and the spaces between them so large, that it is very unlikely that any will physically collide when the galaxies come together (we will return to this idea in chapter 5).

The other galaxies living in our Local Group are smaller than the Milky Way. The largest of these is Triangulum, a galaxy a little further away from us than Andromeda. Triangulum is not in fact triangular, but disc-shaped, owing its name to its presence in the sky within the bright triangle of Milky Way stars that form the Triangulum constellation. Triangulum appears in Messier's catalogue too, as M33 to Andromeda's M31. It is Messier's catalogue names that are now most commonly used by astronomers. Of the other known galaxies in our Local Group, perhaps fifty of them are small 'dwarf' galaxies orbiting like satellites around our own Milky Way. Around twenty more are dwarf galaxies orbiting Andromeda. Even now we are continuing to discover new dwarf galaxies orbiting around us.

It was only in the 1920s that astronomers discovered that these neighbours and other nebulae were actually objects outside our Milky Way. This possibility had long been a source of disagreement among astronomers and other intellectuals, most famously voiced in the 'Great Debate' between American astronomers Heber Curtis and Harlow

Shapley. In two lectures delivered on a Monday afternoon in April 1920, the two astronomers argued over these nebulae, and what they meant for our understanding of the scale and nature of the universe. Shapley argued that the Milky Way was all that existed, that is, that the universe was made up of one single galaxy; but Curtis held that some of the nebulae were, in fact, separate galaxies distinct from our own, or 'Island universes', a term he borrowed from the philosopher Immanuel Kant, who had speculated more than a century earlier that the nebulae were outside our Galaxy. Henrietta Leavitt's discovery of the Cepheid star pulsation pattern in 1908 proved to be the key that allowed astronomer Edwin Hubble to resolve this debate in 1924. By finding Cepheid stars in some of these faint smudges of light in the night sky, Hubble worked out that they must be outside the Milky Way. The stars were simply too dim to be in our own Galaxy. We will return to this topic in chapter 4.

If we keep expanding our view outwards from our galaxy group or a galaxy cluster we discover even bigger entities. These are the superclusters, the largest objects we can see in the whole of our universe. Consisting of hundreds of clusters and groups, they are less well defined than single galaxies or even than clusters, as they don't have such apparently clear edges. In fact, astronomers even now do not agree on where our own supercluster starts and finishes.

The collection of clusters, groups and galaxies in a supercluster are held together by gravity; this is what gives each supercluster its loose definition. One way of establishing its edges is to say that a cluster or a lone galaxy belongs to it if it is moving more towards that supercluster's other members than towards another supercluster. But there is not consensus on the precise rules of membership. Other astronomers say we should think of superclusters as the group of objects that will eventually collapse towards each other at some point in the future.

Until 2014, astronomers were mostly in agreement that the entirety of our own supercluster was a collection of about 100 clusters and groups of galaxies called the Virgo supercluster, comprising a total of around a million galaxies and named after the largest component galaxy cluster. Astronomers calculated that, altogether, the Virgo supercluster is about ten times larger across than the Local Group, over 100 million light-years from side to side. It was only in the 1970s and '80s that new surveys of distant galaxies could be used to figure out Virgo's shape. The observations showed that the largest part looks like a squashed oval, like a rugby ball or an American football, with a large galaxy cluster at its centre and long, stringy filaments of smaller galaxy groups around it.

Recently though, a closer look at the motion of galaxies in superclusters surrounding Virgo suggested that the boundary lines of our supercluster should be redrawn much further out. This would make Virgo just a corner of a much larger supercluster, named Laniakea, or 'Immeasurable Heaven' in Hawaiian, that combines Virgo with three other objects previously thought to be superclusters. Laniakea is about five times wider across than Virgo, and has an irregular shape. Our own Local Group would be a mere watermelon in a basketball-court-sized Laniakea. The status of Laniakea as a supercluster is still uncertain, however, with a study in

2015 showing that its member galaxies and clusters will likely move apart from each other in the future.

Laniakea is so large that if we were to look out to its edges, we would be seeing a few hundred million years back in time. The light that is reaching our telescopes today from those distant parts set off before even the dinosaurs were alive. As that light travelled through space, the dinosaurs came into being, lived and, about the time the light reached the edges of the Virgo supercluster, went extinct. It then had about 60 million more years to go until reaching the Milky Way and finally Earth.

To find out the scale of our supercluster, be it Virgo or Laniakea, we have to rely on a new method of measurement, as the most distant Cepheid stars that we can see with powerful telescopes are about 100 million light-years away, only part of the way out through our supercluster. To judge the distance to more distant galaxies we use instead another standard candle: enormously bright exploding stars that just briefly outshine an entire multi-billion-star galaxy. These are stars of a particular type known as white dwarfs, which we will meet in the next chapter. White dwarf stars become unstable when their mass grows larger than 1.4 times the mass of the Sun, this specific mass worked out by the Indian astrophysicist Subrahmanyan Chandrasekhar in 1930. An explosion is thought to take place either when two white dwarf stars orbiting around each other merge together, or when one white dwarf draws in extra mass from a companion star, becoming suddenly unstable. The exact nature of how this happens is still not known, but the explosions, named Type Ia supernovae, follow a similar pattern. They become increasingly bright and then fade over days or weeks. Astronomers have determined that their maximum brightness relates in a predictable way to how long they stay bright. Following the same principle as with the Cepheid stars, we can use the length of time a supernova stays bright to infer its intrinsic brightness, and then, by measuring how bright it appears to us on Earth, we can work out how far away the explosion took place. These supernovae allow us to measure the distance to galaxies as far away as 17 billion light-years, far beyond our supercluster and out into the farthest reaches of the universe.

We now take our final step outwards, arriving at the extraordinary viewpoint that takes in our entire observable universe. On this largest scale the universe appears as an intricate network of galaxy superclusters that together contain about 100 billion galaxies. Those galaxies are themselves huddled together throughout space in their smaller collections of clusters and galaxy groups. Each of those galaxies has around 100 billion stars, and a huge number of those stars will have their own systems of planets orbiting around them. With such numbers, it is no wonder that most astronomers suspect that life exists in some form elsewhere in the cosmos.

When we refer to 'observable' universe we mean what we are able to see from Earth. What limits this is not how good our telescopes are, but how old the universe is. The universe as we know it has not been around for ever. If we are to be able to see some distant galaxy, that means its light has had time to travel through space to us on Earth. A galaxy that is further away, so far away that its light has not yet had time to get to us, is beyond our cosmic horizon, and beyond our reach.

So how far away is this horizon? We will come around later, in chapter 4, to the idea of the birth of the universe and its age. For now we can say that astronomers have worked out that the cosmic horizon is about 50 billion light-years away from us in all directions. It is more than 14 billion light-years, the distance light could travel during what we now know to be the life-span of the universe, because space has been growing during that time. Our observable universe is therefore spherical, centred on ourselves here on Earth. This does not of course mean that we are at the middle of the universe. We are just, by definition, at the middle of the part we can see. If we now imagine putting the whole observable universe in our basketball court, our home supercluster Laniakea would be about the size of a cookie right in the centre.

To find out how far away those very distant galaxies and superclusters are, the ones in superclusters beyond our own, we can still use the bright supernovae. They take us out close to the very edge of what we can observe. But as we look out towards the edges of our observable universe, we have to grapple with a particular feature of astronomy that is both good and bad. This is the time-machine nature of observing distant space. Although our present-day universe is pretty similar everywhere, filled with galaxies, clusters, superclusters, we view it in a slightly unusual way, because of that fact that light takes some time to reach us. We see things in space as they were when the light set off from whatever it is we are looking at. This means that we see the nearest layers of space to us as they were hundreds to thousands of years ago.

Further layers outwards show their million- and then billion-year-old faces to us, when the universe was much younger. Right at the edge of the visible universe we are seeing the very young universe, which turns out to look quite different from the older parts. By now, those distant parts have presumably evolved to resemble our local part of space.

This is both marvellous and a bit confusing. It means we can never see the whole of space as it is right now, today. But it does mean that we can see into the past, and we can see what other parts of space used to look like. This is fantastically helpful to us, as it helps us piece together how we came to be. Imagine if an alien encountered a large crowd of human beings who were all exactly eighty years old. Just by looking at the group the alien would have a hard time knowing how they had started out, how they had been born, and how they had grown up.

Now imagine that instead our visitor saw a group of humans of mixed ages, with some babies, some children, some young adults, some middle-aged people and some in their later years. It would be much easier to form opinions on the trajectory of human life and the nature of its various stages. We get to do something similar when we look deep out into space and see other galaxies, other stars, as they were in the past. This helps us think about how our own part of space, including our Solar System, has changed over millions and billions of years.

We have reached the end of our journey, stepping outwards and outwards from the Earth all the way to the largest objects in space. If you were to write our entire cosmic address, it would go something like this: Earth, Solar System, Solar

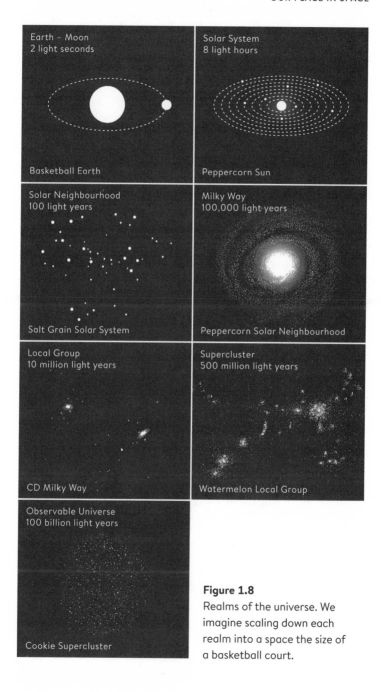

Figure 1.8
Realms of the universe. We imagine scaling down each realm into a space the size of a basketball court.

Neighbourhood, Milky Way, Local Group, Virgo or Laniakea supercluster, observable universe. Just like in looking at an atlas of the world, we avoid trying to picture all those different scales or realms at once. Our brains cannot easily handle the range of sizes and different levels of detail. In an atlas we might look at a map of the world, and after that a map of a whole country, then a region and then perhaps a final view of a town. The same thing is true in astronomy. Each realm of the universe is manageable if we try to limit ourselves to thinking about how it relates to the next one down or up.

Similar to an atlas of the world, we can also imagine starting at the highest level, the map of the whole universe, and stepping down into an entirely different home. Down into a different supercluster, into a different galaxy group, into a different galaxy, different star and landing on a different planet. Our Earth is only one of very many landing destinations.

What does our universe look like beyond the horizon? We think that it continues on, more or less the same, far beyond the part that we can see. We do not think it has an edge, an idea we will come back to in chapter 4. Perhaps the universe is infinite, as hard as that might be to visualize. However large it is, it most likely contains more of the same: more superclusters, containing more galaxies, containing more stars and planets. This might sound superficially boring, but with a moment's thought we can start to imagine what incredible richness and diversity there must be out there. How many galaxies, how many stars, how many planets – and how many of those might be hosts to life?

CHAPTER 2

We Are Made of Stars

In this chapter we will discover more about the stars that light up our sky, from the Sun to those far out beyond our own Galaxy. The origin of heat, light and life, stars are fundamental to our very existence. They produce the basic ingredients of the air we breathe, the food we eat and the cells that make up our bodies. We will find out why they shine, how they live and die, and what other worlds they could be holding around them. To get to know stars in all their glory we first need to understand the workings of the fundamental tools of astronomy: light and telescopes.

Light flies through space at the incredible speed of 700 million miles an hour, and we can think of it moving like a set of waves to reach us, each wave undulating up and down like ripples on a pond after a rock has been dropped in. Ripples on a pond might have a regular spacing between the crests, a wavelength that might be tens of centimetres long. Light with a particular wavelength also has a regular spacing between its crests. The light that our own human eyes have evolved to see is of a very particular sort, made up of all the colours of the rainbow from red to violet. Our eyes interpret different wavelengths as different colours, and the wavelengths we can see are much shorter than the waves on

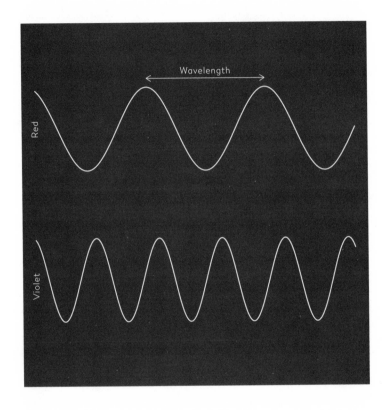

Figure 2.1
The colour of light depends on its wavelength.

the surface of the pond. The colour red is the longest wavelength we can see, with a tiny spacing between its crests of just less than a thousandth of a millimetre. After that come orange, yellow, green and blue, each one a little shorter than the last. Violet has the shortest wavelength of all, at about half of red's wavelength.

When we see white or yellowish light from a lamp or the Sun, it is made up of many waves of different wavelengths, spanning all the colours of the rainbow. It was Isaac Newton who first showed this to be true in 1672, experimenting by bending light through prisms of glass. Light travels more slowly in glass, or in water, than it does in air, and the shorter the wavelength of the light, the more it gets slowed down. So, blue light travels slower than red through the glass, and the effect of this is to bend the blue light as it enters the glass more than the red. The light entering the prism is white, but it emerges split up into the rainbow of colours. We can see the same effect in a rainbow in the sky too: there the sunlight is bending through many droplets of water, separating it out into all of its visible wavelengths.

This rainbow of visible colours is the most familiar to us, and our eyes likely evolved to see these particular wavelengths of light because they are the kind that our Sun sends out most plentifully. But what we can see is only a fraction of light's fuller nature. Light can also have wavelengths with crests much further apart or much closer together. These are forms of light such as infrared light, ultraviolet light and radio waves, which our eyes cannot see.

To do astronomy we look at light coming from objects in the sky. But our eyes, miraculous as they are, have two obvious limitations. We cannot zoom closer to see the source of light, and we cannot see the non-visible light. Astronomers have long tried to compensate for these limitations, and the key to addressing them is the vast range of telescopes that we have developed over the past 400 years.

The earliest telescopes were of the kind we know best: refracting telescopes that use a glass lens to focus rays of visible light. Light bends its path when it enters the curved concave-shaped lens at an angle, because the glass slows down its speed. A lens can take many rays of light and, following its curved design, bend them towards a smaller area. We can then use a second, smaller eyepiece lens to bend the light onto a straight path into our eyes. This act of focusing many rays of light magnifies and brightens the image. The larger the first 'objective' lens, the more light it focuses. The first telescopes appeared in the Netherlands in the early 1600s from the eyeglass makers Hans Lippershey, Jacob Metius and Zacharias Janssen, who discovered that two eyeglass lenses could be used together to magnify images. These 'lookers', as they were originally known, were eagerly adopted by the Dutch army to watch for enemy boats at sea. The idea also spread to Galileo in Venice, where, in 1609, he made a telescope of his own using a lens 4 centimetres across to magnify objects about twenty times. As we have learned already, he famously turned it to the sky, finding craters on the Moon and discovering the large moons of Jupiter. He could see stars that were not visible to the naked eye and he referred to the largest and brightest of them as 'seventh magnitude': fainter than all the stars that had been observed and classified before.

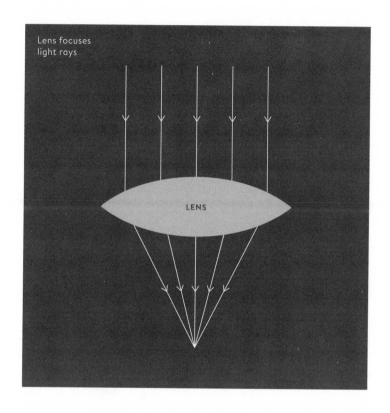

Figure 2.2
A lens changes the direction of light rays.

In 1611 Johannes Kepler refined Galileo's telescope design. Galileo had used a convex-shaped eyepiece lens, and Kepler changed its shape to become concave. He then positioned the eyepiece just far enough away from the objective lens so that all the rays of light bent through the objective lens, focused to a point, and then continued before hitting the eyepiece. The image you see with this design is upside-down, but you can also see more of the sky at once. Refracting telescopes using Kepler's design remain common in amateur astronomy and are vital in many professional settings too. The largest refracting telescope still in use is at the Yerkes Observatory in Wisconsin, with a lens a metre across. Instead of looking through them with our eyes, the images for professional use are now recorded with digital cameras.

Refracting telescopes cannot get much larger than a metre because gravity would deform the glass. To get around this limitation, the largest professional telescopes are reflecting telescopes. Instead of a lens, they use a curved mirror to collect the light, reflecting it onto an eyepiece or camera at a smaller focus point. These telescopes can be much larger than the refracting type, with the largest mirrors giving the greatest resolution of fine details. One of the earliest versions of these was developed by Isaac Newton in 1668, using a mirror just over an inch across.

These reflecting telescopes revolutionized our ability to see far into space, and the largest now have mirrors of 8 to 10 metres across. Telescopes of this scale allow us to see objects as faint as twenty-seventh magnitude: this is twenty-one steps in magnitude fainter than we can see with our naked eyes on a dark night, which works out to be an incredible few hundred million times fainter. To see distant objects in ever-better detail, a team of astronomers are constructing the Giant Magellan Telescope in Chile, which will combine a set of seven 8-metre-sized mirror segments to give a reflecting mirror more than 20 metres across when it begins operation in the 2020s. At the same time the European Southern Observatory is building what is aptly called the 'Extremely Large Telescope', which will have a mirror 40 metres across.

Most telescopes operate here on Earth, so the quality of the images we can capture is limited by having to see through our atmosphere. The same effect that causes stars to twinkle makes magnified images of stars or galaxies blurry. This problem is partly overcome by locating telescopes in elevated places on Earth where there is less atmosphere to see through, and selecting sites where it is dry and cloudless and the air is still. The best places for this include the Maunakea volcano in Hawaii, mountaintops in Chile and volcanoes on the Canary Islands.

A telescope in space can see even more clearly. Astronomers had ambitions to do this as early as the 1920s, but it wasn't until 1990 that the famed Hubble Space Telescope was launched into Earth's orbit, a reflecting telescope with a mirror almost 2.5 metres across. The road to put the telescope into space was long, spearheaded by American astronomer Lyman Spitzer, who proposed the idea of a large space telescope in 1946. By the 1960s the astronomical community was convinced of its importance, and by 1970 the project was supported by NASA. It almost didn't happen though: the United States Congress cancelled funding in 1974 due to budget cuts. Intense lobbying by astronomers, led by Lyman Spitzer and astrophysicist John Bahcall, both based in Princeton, led to its approval for funding a few years later. After launch in 1990, Hubble's challenges were not yet over: when it began sending back images, it became obvious from their poor quality that the mirror had not been ground correctly. NASA astronauts repaired it during a heroic mission in 1993, readying it for more than two decades of fantastic service to the astronomical community. It has given us perhaps the most iconic pictures in astronomy.

Until the 1930s, it was only visible light that telescopes magnified and collected. Now, we are able to see the universe in its full glory, using telescopes that measure all the other sorts of light. Many of these are only visible from space, unable to penetrate our Earth's atmosphere. We will now step through those different types, moving out first through longer waves than we are able to see, then in the other direction to shorter waves.

Light that has crests just a bit longer than our eyes can see is called **infrared light**. About half of the energy that the Sun sends to Earth arrives in this form. Infrared light is also produced by anything that is hot, including our own bodies, which comes in useful for professionals like firefighters, who use infrared cameras to look for warm bodies in smoke-filled buildings. Some animals – among them snakes, piranha and mosquitoes – can sense infrared light with their eyes or with other parts of their bodies, something we can do to some extent by sensing heat with our skin.

William Herschel first discovered infrared light, in 1800, while using a prism of glass to split up light into the rainbow of its visible colours. He was interested in measuring the temperature of each of the colours, placing a thermometer in each part of the rainbow. He found that the red light was hotter than the violet light, and as part of his experiment he moved his thermometer just past the red part of the rainbow. Unexpectedly, he found that the temperature there was warmer than in all the other parts he had measured. He worked out that he must have found a new kind of light that was warm and invisible to our eyes.

Telescopes designed to measure this infrared light are particularly good at seeing things in space that are warm but not blazing with visible light. They work in the same way as optical telescopes, but instead of directing the light into an eye or an optical camera with CCD detectors, they direct it at cameras with detectors designed to pick up the longer wavelengths of infrared light. In fact, most cameras on our mobile phones are similarly attuned to see some infrared light. You can watch this in action by pointing a remote control at a phone and taking a picture while pressing a button. Remote controls send a beam of infrared light to transmit instructions, so when a button is pressed the infrared light should make a dot appear on the camera.

In astronomy, infrared light is especially important as it can get through matter that blocks visible light, like clouds of cosmic gas and dust, so it allows us to peer into parts of space that would usually be hidden. Our atmosphere, however, thick with water vapour, carbon dioxide and methane, blocks most of the infrared light coming from space, so the best infrared telescopes operate out in space, beyond our atmosphere.

Continuing on the spectrum of non-visible light, **microwaves** have a longer wavelength than infrared light, ranging from around a millimetre up to about 10 centimetres. This is perhaps the most tangible of light waves, thanks in part to its association with the common microwave oven. These ovens work by filling an enclosed space with microwave light, which bounces back and forth off the walls. The light makes the water molecules in the food rotate and collide with other nearby molecules, which heats it up. There is a fun experiment that can be done with an old-fashioned microwave to see the pattern of waves in the light. If you put a paper plate of chocolate chips in the microwave, remove any rotating device and turn it on for a few seconds, you should see that some chips are largely unaffected while others melt in spots separated by about 6 centimetres. Those hot spots mark the crests in the microwave light waves travelling to and fro across the oven, where the light is most intense. The reason for spinning food around in a microwave is to prevent the same part of it sitting in the hottest spots all the time. The experiment won't work in a microwave designed to operate without a spinning plate, as the source of microwave light in those is being rotated instead, and the chips will melt evenly.

We use microwave light in many ways beyond heating food. When you connect a device with wi-fi, for example, microwaves travel between the nearest router and your phone or laptop, carrying information back and forth and connecting you to the world. The abundance of man-made microwave noise can make it hard for astronomers to distinguish objects in the night sky that also send out faint microwave light. We have to design microwave telescopes carefully to avoid being drowned out by our own signals. An important astronomical signal of microwave light comes from the most distant part of space we are able to see; we will get to this in chapter 4.

The very longest wavelengths of light have undulations from about 10 centimetres long up to many miles. These are **radio waves** and, like microwaves, they are everywhere. They are flying past us and through us all the time. Radio waves can travel through walls, which is why your radio and television can pick up radio-light signals inside your house and turn them into sounds and pictures. As well as carrying signals for radio and television stations, we use radio waves constantly in modern communication. Our mobile phones, notably, send and receive wavelengths that are about a foot long. When we speak into a phone, the radio light encoding our words is sent from our phone to a succession of phone masts scattered across the country, and then on to the receiving phone of the person we are speaking to. The signal takes only milliseconds to get from one person to the other, so we can talk almost in real time.

Astronomers have conducted a wealth of fascinating research by examining radio waves coming from space, including discovering and studying rapidly spinning stars and seeing rotating discs of matter drawn into giant black holes at the cores of galaxies, often sending out huge jets of radio light. We will learn more about these in this and the next chapter. Our atmosphere allows through radio light because their wavelengths are much longer than the size of molecules in the atmosphere, so they don't scatter or absorb the light. This means we can build radio telescopes here on Earth, and we can use them during the daytime too. These telescopes usually have much larger dishes than optical or infrared telescopes, because the longer wavelength of the light means that a larger mirror is needed to see the same degree of detail. The dishes reflect the radio light onto a radio receiver, the equivalent of a camera for visible light. Some of the best-known radio telescopes are the 300-metre Arecibo telescope in Puerto Rico, built in a natural depression in the land, and the 100-metre Green Bank and Effelsberg telescopes in West Virginia and in Germany. Jodrell Bank's Lovell Telescope in the UK is almost as large, at 80 metres across.

As with microwaves, a major challenge for astronomers is to find locations on Earth where the interference from human use of radio waves for communication is minimal. At the moment, two of the best places are in the dry South African Karoo desert and the desert-like plains of Western Australia, both inhospitable to humans and, as a result, fantastically radio-quiet. A vast network of radio telescopes known as the Square Kilometre Array is now being built in both places, with many small dishes spaced widely apart. By connecting them together into an array of telescopes, which means using computers to keep track of which signals are arriving at all the small telescopes at any given time, it will collect as much light as a telescope an entire kilometre across and will see with even greater detail than a single dish. Even more ambitious plans imagine a futuristic radio telescope on the far side of the Moon, a place certainly well shielded from human chatter.

Turning now to the waves of light shorter than our rainbow of visible light, we come first to **ultraviolet**, hovering just beyond our eyes' reach but visible to bumblebees and various other insects and birds. Its wavelengths are so short that they are hard to visualize, reaching as small as millionths of a millimetre. We know ultraviolet light best as the harmful part of the Sun's light that, in excess, can damage our skin. Fortunately, most of it gets blocked by the ozone in our atmosphere.

Johann Ritter, a German chemist and physicist, discovered ultraviolet light in 1801. Inspired by Herschel's recent discovery of infrared light, he decided to investigate the other end of the rainbow. He experimented with the chemical silver chloride, which turns black only when exposed to light, and discovered that it turned black fastest when placed just beyond the violet end of the rainbow. This only made sense if there was some invisible light there too, and light that was more energetic than the visible violet light. Light with a shorter wavelength has more crests every second and more energy. The ultraviolet light has enough energy that it can damage the DNA inside skin cells, leading to mutations that cause cells to grow unregulated.

Like our Sun, other stars shine brightly with ultraviolet light, some of them emitting even more of it than visible light. Telescopes equipped with ultraviolet cameras have taken extraordinary pictures of stars and galaxies in deep space. The Hubble Space Telescope, for example, used not just visible light but also ultraviolet and infrared light to make many of its images.

X-rays, with shorter wavelengths than ultraviolet light and even more energy, come next on the spectrum. We are familiar with this kind of light from hospital scans and

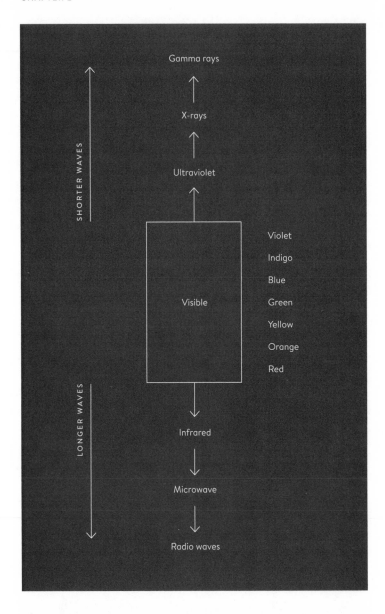

Figure 2.3
The full range of light, which includes the visible rainbow as well as
light with longer and shorter wavelengths.

airport security. X-ray light will more likely be absorbed by an atom that has a large number of electrons, since the process involves taking in some X-ray light and ejecting an electron. Most of our bodies' atoms, like carbon, have nuclei with too few electrons, so the light travels through the soft parts. It stops only when it meets bone, since the larger calcium atoms do absorb X-ray light. X-ray photographs are produced by shining X-rays through our bodies onto film that turns black where the light hits it, but stays white where the bones have blocked the light. It was German physicist Wilhelm Röntgen who invented X-ray photographs in 1895, taking the very first X-ray picture, of his wife Anna's hand. He called them X-rays, with the 'X' signifying something new and unknown. The name has stuck in the English language, but in German they are known as Röntgen rays. The medical community instantly began using them, and the discovery won Röntgen the very first Nobel Prize in Physics, in 1901.

Similar to UV light, X-rays are so energetic that excessive exposure can damage the genetic material inside our cells, leading to cancer. This is why in hospitals we use them sparingly. For astronomers, their high energy means that they are produced from hot gas that has been heated up to millions of degrees. This is found throughout the universe, as we will discover in the next two chapters, and being able to see it with X-ray eyes allows astronomers to study extreme objects and events, including exploding stars, colliding galaxies, black holes and giant galaxy clusters. Earth's atmosphere, however, absorbs X-rays, so the only way to see the X-rays from astrophysical objects is to send telescopes above our atmosphere either aboard a satellite or a high-floating balloon. Many of our recent observations have come from NASA's Chandra X-ray Observatory satellite, observing the sky since the late 1990s.

Beyond X-rays, we find **gamma rays**, the most energetic of all forms of light, with a wavelength about the size of an atom. It is even more lethal than X-rays in high doses, though its ability to break up cells in our bodies does make it useful for certain focused medical treatments. Gamma-ray telescopes, which usually have to be launched beyond the atmosphere, allow us to see the most energetic things happening out in our Galaxy, and even more so in distant galaxies. Many of these gamma-ray signals are produced by conditions of intense gravity and strong magnetic fields that we are still seeking to understand, including dense neutron stars, exploding stars and matter swirling around black holes. NASA's Fermi Gamma-ray Space Telescope launched in 2008 and is surveying the gamma-ray sky to study these phenomena. Bursts of gamma-ray light from beyond our Milky Way are regularly seen by the Fermi telescope and by NASA's Neil Gehrels Swift Observatory, launched in 2004. They likely come from collapsing or colliding stars, but their cosmic source is still not fully known.

From radio waves to gamma rays, the diversity of light, and the variety of telescopes and other instruments we have created to see across its full range, allow us to explore space in ways that we could never do with our eyes alone. And as soon as we developed the first telescopes, astronomers trained them on those objects that had for so long been such a source of wonder: the stars. These earliest telescopes helped astronomers identify stars, construct maps and catalogues, and measure their brightness and colours.

Now we can do more than that: we can better understand the inner workings of stars. For the past century we have understood that stars are huge balls of gas made up mostly of hydrogen and helium. Like Jupiter, stars have no hard surface you could stand on, even if you could withstand the heat. For much of their lives, stars are kept in shape by a delicate balance between gravity's pull, drawing the gas inwards, and pressure from the hot gas that pushes it back outwards. The pressure arises from the gas having a high temperature, making for fast-moving particles that push against each other and resist being squeezed together. We can see gas pressure in action on a small scale when a covered pan of water starts to boil and tries to push its lid off. In a star the source of that pressure is far more intense, and it comes from extraordinary activity taking place right at its very centre.

There at the core of every star is something a little like a nuclear power station, squeezing pairs of hydrogen atoms together to turn them into larger atoms of helium and releasing immense amounts of energy in the process. This is nuclear fusion, partner to the nuclear fission that we currently use to create energy in power stations. In fission, energy gets released when large atoms of uranium are broken up into smaller ones. Because fusion uses hydrogen, which is abundant, and does not produce radioactive waste products, scientists have long dreamed of using it to create an alternative and sustainable source of energy. But fusion is hard to start and maintain, and no one has yet been able to successfully control it. Fusion's only man-made use so far has been in nuclear bombs.

To make fusion happen, to squeeze atoms together, requires extraordinary compression. In a huge star, this is provided by gravity. Remember that our own Sun is a hundred times the size of the Earth across, which makes it a million times larger in volume. It is not as dense as the Earth but it still weighs 300,000 times as much as our planet. This is massive enough to make the gravity so intense and the core of the star so hot and dense that the hydrogen atoms ignite. You need a star with at least a tenth of the mass of our Sun for this to happen, with temperatures reaching a few million degrees.

Once things get hot enough in the middle of a ball of hydrogen and helium gas, fusion of the hydrogen gets going in earnest, and the star as we know it is born. The energy created by the fusion pours out as heat and light and presses the gas outwards. In our Sun, as in other stars, the fusion only happens at the core, in a space that would be the size of a golf ball if the Sun were as big as a basketball. If the light could travel freely, it would escape from the star in seconds. Instead, the light has to pass through dense material to get out, so it keeps changing direction as it hits atoms in its path, taking tens of thousands of years to leave the star. But once free, it is just a speedy eight minutes in a straight line all the way to Earth.

It was only in the 1920s that we learned what stars are made of, and in the 1930s that we understood fusion to be the source of their light. Much of our understanding is thanks to Cecilia Payne-Gaposchkin's pioneering work on the spectra of stars. Payne-Gaposchkin was inspired to take up

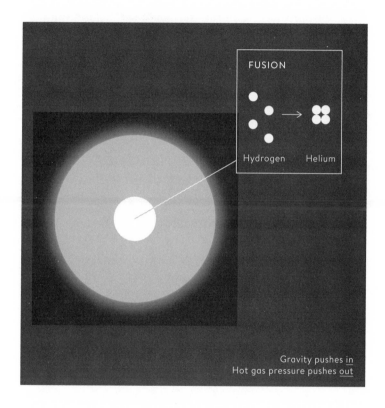

Figure 2.4
A star makes heat and light from fusion at its core.

astronomy as a student at the University of Cambridge when she heard a talk by Arthur Eddington about his 1919 expedition to observe the stars near a solar eclipse, a journey that famously confirmed Einstein's new theory of gravity. Unfortunately, the opportunities to work as an astronomer in England were limited; Cambridge did not even award women degrees at the time. So, in 1923, supported by Harlow Shapley, she sailed to the other Cambridge, in Massachusetts, to study for a PhD at the Harvard College Observatory. She would remain at Harvard for the rest of her illustrious career, where in time she became the first woman professor in the Department of Astronomy and the first woman to chair any department at Harvard.

Harvard was an extraordinary place for astronomy in the 1920s, not least because of the 'Harvard Computers', the group of women who had been working for astronomer Edward Pickering since the late nineteenth century and which included Henrietta Leavitt of Cepheid fame. Only men were allowed to operate the observatory's telescopes, so the women studied, analysed and catalogued data and photographs. The women were paid vastly less than the men for their work, yet they managed to make many of the most exciting discoveries.

One of Pickering's goals was to measure the spectra of as many stars as possible, and to use these to classify the stars into types. To measure the spectrum of light means to spread that light out into the rainbow and find out how intense the light is in each colour, for example how much of it is red, orange, yellow, green, blue. Astronomers had noticed since the nineteenth century that when we examine the spectrum of starlight, we see a rainbow of colours, but with dark gaps in the rainbow where certain colours are missing. These gaps correspond with particular wavelengths of light that have been absorbed by atoms in the atmosphere of the star as they beam out from the core. Different gases in the atmosphere of the stars absorb light of different wavelengths.

Pickering's group published its first catalogue, the *Draper Catalogue of Stellar Spectra*, featuring more than 10,000 stars and their spectra, in 1890. It was named for Henry Draper, a doctor and amateur astronomer, who had measured some of the first stellar spectra in the late 1800s. Draper's widow and collaborator, Mary Anna Draper, had taken an interest in Pickering's work and funded his project to produce the first comprehensive catalogue of spectra. The catalogue used letters running from A to Q to categorize the different spectra of stars, a scheme pioneered and implemented by Williamina Fleming, another Harvard 'Computer' – and one with a very unusual career path. In 1878, twenty-one-year-old Fleming had emigrated from Scotland to Boston with her husband and child, but her husband abandoned her when they arrived. Pickering employed her first as a maid in his household, but, recognizing her abilities, he then asked her to work at the observatory after reportedly becoming unhappy with the progress made by his male assistants. She arrived and began to develop her system, according to which, 'A' stars were those which appeared to have the most hydrogen in their atmosphere, creating the darkest gaps in the rainbow; 'B' stars had a little less hydrogen and so on. Some of the other letters were assigned to represent other elements thought to be in the stars' atmospheres.

Yet another member of the Harvard Computers, Annie Jump Cannon, made important adjustments to Fleming's system. Like Henrietta Leavitt, Cannon had become almost deaf in early adulthood, but she devoted herself to her work, joining the Harvard College Observatory in 1896 and classifying about 350,000 stars during her life. Her breakthrough in 1901 was to use a simpler ordering of the stars than Fleming, reordering them by colour from blue through to red, and dividing them up into only seven categories, using Fleming's letters O, B, A, F, G, K and M. The order has long been remembered since Cannon's time by the widely used but now rather dated mnemonic 'Oh Be A Fine Girl, Kiss Me'; in modern usage 'Girl' and 'Guy' are used interchangeably. The International Astronomical Union officially adopted her classification system in 1922, and it is still used today.

With the stars divided into classes, astronomers began to look at the pattern relating the spectral class, or colour, of a star to its intrinsic brightness. An early example of this comes from German astronomer Hans Rosenberg in 1910. He took forty-one stars in the Pleiades star cluster, all at the same distance from Earth, and found that for most of the stars, the bluer they were, the brighter they were. Danish astronomer Ejnar Hertzsprung published a paper in 1911 that looked at the pattern for stars in both the Pleiades and Hyades star clusters, and in 1912, one of the key figures in the field at the time, Princeton astronomer Henry Norris Russell, presented a version to the Royal Astronomical Society that included more stars. They both found the same trend that, for most of the stars, the bluer the star, the brighter it was. These stars were referred to by Hertzsprung as 'dwarfs'.

But a number of them did not follow this pattern: some of the redder stars were particularly bright, about 100 times as bright as the Sun, and were dubbed 'giant' stars. This way of looking at the stars' colour and brightness became known as the Hertzsprung–Russell diagram, still in common use in astronomy.

At this point, the stars were well classified, but no one had yet worked out what they were made of, or why they followed these particular trends relating their colour to their brightness. Astronomers at the time saw patterns in the stars' spectra that were the fingerprints of calcium and iron and speculated that stars were probably made of the same mix of elements that make everything here on Earth.

They were quite wrong, and it was Payne-Gaposchkin who first realized it. Armed with an understanding of the new theory of quantum mechanics, and drawing on vital work done by Bangladeshi astronomer Meghnad Saha, she examined Cannon's detailed classifications and concluded that the variation in absorption patterns did not arise, as others thought, because different stars consisted of different elements. Rather, she argued that the main ingredients in all stars were hydrogen and helium and that the variation simply arose from the fact that stars had different temperatures. Cannon's lettering system from 'O' to 'M' corresponded not only to variation in colour, but also to the temperature of the star decreasing from hottest to coolest. The stars were nothing like our Earth, it turned out, and contained only tiny traces of the other elements heavier than helium. Payne-Gaposchkin's conclusions were initially met with heavy scepticism. Henry Norris Russell discouraged her from presenting her new results in her doctoral dissertation in 1925, as they defied conventional wisdom. But Payne-Gaposchkin had got it right, and Russell came around to agreeing with her a few years later. Her work, after thousands of years of astronomical speculation and research, had finally revealed what stars are made of.

Before Payne-Gaposchkin had made this great advance, the well-known British astronomer Arthur Eddington had already speculated in 1920 in *The Internal Constitution of the Stars* that the source of the star's energy might be coming from fusion of hydrogen. Using Einstein's theory of relativity, which said mass could be turned into energy, he worked out that if just 5 per cent of the star's mass were hydrogen, and it were fusing to make heavier elements, it would produce enough heat to explain the observed starlight. This idea turned out to be correct. Astronomers would soon find out that even though most of the star was made of hydrogen, only the central core was hot and dense enough to be fusing atoms together. The theory was worked out in detail by German-American physicist Hans Bethe in the 1930s, culminating in his 1939 paper titled 'Energy Production in Stars'. It would win him the 1967 Nobel Prize in Physics and explained the trend of the 'dwarf' stars: more massive stars have a stronger pull of gravity at their cores and so are undergoing more intense fusion, making them both brighter and hotter.

Stars may all be balls of mostly hydrogen gas, but they lead very different lives depending on how massive they are when they are born. We can look at their different life stories by grouping stars according to their colour at the start of their lives, which tells us how heavy they were. Rather than think about all seven classes, we will broadly group them into four colours – red, yellow, white and blue – that range from the least to most massive. The majority of stars – about 90 per cent of them – are red. They are the coldest and weigh the least, with temperatures at their surfaces ranging from 'only' 3,000 to 5,000 degrees. A little more massive are the yellow stars, of which our own Sun is one, that together make up around 10 per cent of all the stars we know. They have temperatures between 5,000 and 8,000 degrees at their surfaces. Hotter and heavier than the yellow stars are white stars, much less abundant – only about one in every 100 stars. The rarest and hottest of all are blue stars, perhaps one in 1,000 stars, and with surface temperatures upwards of 25,000 degrees.

We will start with the life-story of our Sun, the story of the **yellow stars**. For much of its life, our Sun will remain in its current and familiar form. Right now, it is about halfway through that stage of its life, which we expect to last about 10 billion years. For all this time it will be supported by the delicate balance between its inward-squeezing gravity and the outward-pushing pressure produced by the heat created from fusing the hydrogen atoms at its core. The temperature on the Sun's surface is a fiery 6,000 degrees, and right at the middle reaches as hot as 15 million degrees where the fusion is taking place. During this part of its life the Sun sends out light in all the colours of the rainbow that add up to look whitish, but the colour it produces the most is yellow, due to its temperature at the surface.

The delicate balance that supports the Sun will get

disrupted when it runs out of hydrogen atoms at its centre. We can work out when this will happen by using the mass of the Sun to tell us how much hydrogen 'fuel' it contains, and finding out how fast it is burning through the hydrogen by measuring how bright the Sun is. Putting these numbers together gives us an estimate of 5 billion years from now. When this happens, once the hydrogen atoms in the core have all turned into helium, the temperature will not yet be hot enough to fuse those atoms together to form slightly bigger ones, like carbon and oxygen. So instead, for a while, gravity will win out and pull the gas at the core of the star inwards, squeezing it and making it hotter and hotter until the hydrogen atoms around the core get hot enough to burn.

At that point the outer layers of our Sun will swell up dramatically. This will happen because the outer layer has a greater volume of hydrogen and so produces more outward-pushing pressure, increasing the volume of the star. It will grow hundreds of times larger from side to side. It will get much brighter as the temperature at the core rises, but it will also be spreading its heat over a much larger surface so will start to glow a cooler orangey-red. A new phase of the Sun's life will begin, life as a red giant.

This should happen about 5 billion years from now, and it will make the Solar System a very different place. The Sun is likely to grow so large that the orbits of Mercury and Venus will get swallowed up, and probably the Earth's too. If not, we will be perilously close to the edge of our new giant Sun, and conditions will likely be impossibly hot for life, or at least life as we know it. The increased pressure during the red giant phase will also cause the outer layers of our star to

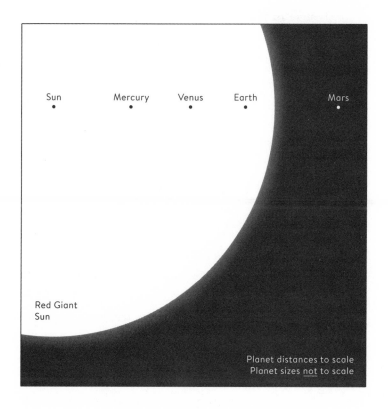

Figure 2.5
The expected size of the Sun when it becomes a red giant.

be gradually ejected in expanding shells of gas. These have long been observed around other stars in our Milky Way and were named 'planetary nebulae' in the 1780s by William Herschel. The gases produce rings of beautiful colours, which were thought to resemble planets.

We expect the Sun to live like that for another billion years. As the hydrogen runs out, the core will collapse further as gravity keeps winning. Finally, when the Sun reaches about 100 million degrees at its core, it will start making carbon and oxygen atoms from the helium. The nucleus of a helium atom has two protons and two neutrons, which fuse to make carbon with six protons and six neutrons, and oxygen with eight protons and eight neutrons. When the helium runs out, the Sun will not be big enough or hot enough to ever fuse carbon and oxygen into anything larger. This will be the end of the line.

Left behind will be a small core known as a white dwarf, made mostly of carbon and oxygen. It is no longer creating any outward-pushing pressure from fusion, so we might expect gravity to win the battle and cause the white dwarf to collapse completely. This doesn't happen, though, and it is due to an elegant example of quantum mechanics at work. Quantum mechanics tells us that tiny electron particles must avoid being in exactly the same place as each other, so if you push a swarm of electrons together, they will resist and push back. There are plenty of electrons at the core of the star, as each of the newly made carbon and oxygen atoms have a few. These electrons create a new outward-pushing pressure that balances the stronger inward-pushing gravity. The Sun has now entered the final phase of its life. It will be extremely dense, weighing not much less than it does now, but extraordinarily squeezed into a space the size of our Earth, a million times smaller in volume than it is now.

In chapter 1 we encountered the idea that a white dwarf star can only ever be as heavy as 1.4 times the mass of our Sun. This is the largest inward-pulling mass that the outward-pushing quantum pressure can manage to balance and it is known as the Chandrasekhar limit, named after the brilliant Indian-American astronomer Subrahmanyan Chandrasekhar, who worked at Cambridge in the 1930s. In fact, it was on his voyage from India to Cambridge in 1930 to start life as a graduate student that he first worked out this limit, having been inspired by early work on the topic by the Cambridge astronomer Ralph Fowler. Chandrasekhar later spent his career at the University of Chicago, after experiencing some unpleasant encounters during his time in Britain, most notably with Arthur Eddington. Eddington thought Chandrasekhar's limit was quite wrong and dismissed it publicly as 'stellar buffoonery' in a meeting of the Royal Astronomical Society in January 1935. Chandrasekhar would later be proved right and was eventually awarded the Nobel Prize in Physics for his achievements, in 1983.

As a white dwarf, the Sun will start out hot, but it will no longer produce any new energy of its own, so it will gradually cool down until it sends out no more light at all. Astronomers predict that it will end its life billions of years in the future as a black dwarf, a cold, dark version of the white dwarf. The universe is still too young to have produced any black dwarfs yet, so we can only guess at their future existence based on our understanding of how stars cool down.

Looking beyond our Sun, our understanding is that all lone yellow stars will lead very similar lives. They will all live roughly 10 billion years in their first phase, transform into red giants and then finally will end their active lives as fading white dwarfs, having thrown off their outer layers of gas into space. We expect this to be true whether these stars are in our own Galaxy or in distant galaxies billions of light-years away from us. Life is a little less predictable for yellow stars that live in a pair orbiting around each other: a binary pair of yellow stars might end up as two white dwarfs which eventually collide.

The majority of stars in the universe have about half our Sun's mass or less, down to about a tenth of its mass. These are the plentiful cooler **red stars**, and they lead their lives much like yellow stars but at a slower pace. They spend the main phase of their life with the same balance of inward-pulling gravity and outward-pressure from hydrogen fusion, but the gravity is not as intense at their centres, as their mass is lower, so the fusion takes place at a more leisurely speed. These stars will live many tens of billions of years before the hydrogen runs out at their cores, so none have made it to that life-change yet. Far in the future those red stars will, again like yellow stars, also become giants and will shrink their cores. They will likely never get hot enough to create carbon or oxygen atoms, however, and will end their lives as white dwarfs, and later black dwarfs, made of helium.

It is the big stars that lead the most extreme lives, living fast and dying young. Weighing from about eight times our Sun up to hundreds of times, these are the rarer **white** and **blue stars**. In our Milky Way galaxy of a hundred billion stars, only about a billion are these stellar heavyweights. A heavier star feels a stronger pull of gravity and has hotter temperatures, burning hydrogen gas at the heart of the star much faster. The smaller of these, the white stars, are from eight to about twenty times heavier than our Sun. Their colour is whiter than our sunlight, as they send out less reddish light than our Sun, and more bluish and ultraviolet light. White stars typically take less than a billion years to burn through their hydrogen, still a long time but about ten times quicker than our Sun. If our Sun had been born bigger, as a white star like Sirius, its life would already be over.

The blue stars are heavier still and can weigh a hundred times as much as the Sun. They are even hotter than the white stars and glow mostly with blue to ultraviolet light. Blue stars live extremely fast, burning through their hydrogen supply in only around 10 million years. This is much less than the time since dinosaurs were roaming around on Earth!

When white or blue stars burn through their core hydrogen supply, they follow the same course as yellow stars, but at high speed. The star's core shrinks, and the outer part expands to turn the star into a huge red giant. Just like with yellow stars, the hydrogen supply just around the inner core starts burning instead, and in these massive stars it is quickly hot enough for the helium in the core to start fusing into the heavier elements like carbon and oxygen.

The final stage of life for these large white and blue stars is spectacularly different from that facing our Sun. The gravity at the heart of the star is so strong that it squeezes the core of the star enough to form not only carbon and oxygen, but also even heavier elements all the way up to iron. It will end up with layers like an onion, with hydrogen nearest the surface, then helium, carbon and the heavier elements up to iron at the core. But when it produces iron, calamity occurs, as fusing iron atoms together releases no new energy. Nature has made iron a tipping point, in that it and all heavier atoms only release energy when broken up, not when combined together.

At that point, when iron is produced, the star is incredibly hot, but suddenly there is no more fuel that can be burned at the core, and the outward-pushing pressure comes to a halt. The great force of inward-pulling gravity takes over. For our Sun at that stage, the pull of gravity will get balanced by that quantum-mechanical pressure from the tiny electrons, which will create a white dwarf. But that process only works for cores of stars that are lighter than the Chandrasekhar limit, 1.4 times the mass of our Sun. This corresponds to stars that started off about eight times the mass of the Sun or less, before they lost their outer envelopes. The mass of the leftover cores of these white stars is higher than that threshold and is too big for the pressure of the electrons to compete with the pull of gravity. The result is an extreme implosion. The inner core totally collapses, followed quickly by the surrounding layers of the star. It happens so fast that the middle of the star rapidly gets as hot as 100 billion degrees.

The implosion stops dead in its tracks when the core gets as dense as an atom's nucleus. At that point it will consist mainly of neutron particles, and a new outward pushing pressure will take over, arising from the 'strong' force. This is a fundamental force that usually acts to hold neutrons and protons together inside atoms but it becomes repulsive when neutrons or protons get too close together and works to push them apart. The sudden halt reverses the inward-implosion of the star and turns it instantly into a massive explosion that hurls the entire outer part of the star into space.

This extraordinary event is a supernova. After living for as much as a billion years, the star explodes in just minutes. The explosion produces a vast amount of light, so much that it can briefly outshine an entire multi-billion-star galaxy. The light from a supernova can be seen for weeks or months afterwards. The light comes out in many forms, from radio waves through visible light to X-rays and gamma rays, so astronomers eagerly look at supernovae using different telescopes sensitive to all the different wavelengths of light. We cannot see inside the star while it all happens, however, so much of what we think is happening is speculation, derived from theoretical ideas that we then test by comparing predictions to the light we see coming out.

These supernovae that come from exploding massive stars are distinct from the 'Type Ia' sort that we met earlier, which are thought to come from white dwarf stars acquiring too much mass to stay stable. There is probably a supernova of some sort happening roughly every 100 years in a galaxy like the Milky Way, but they are surprisingly hard to observe directly. The light from many of them gets blocked by dust and gas in our Galaxy. And we will be missing out on seeing many of those in galaxies beyond our own, as we cannot train our telescopes on every galaxy each night.

If a supernova happens in a galaxy near enough to us, we might be lucky enough to have a picture of the galaxy before and after the event, with the supernova suddenly showing up as a bright spot. In January 2008, Carnegie-Princeton research fellows Alicia Soderberg and Edo Berger were even luckier: they were observing a galaxy while a supernova exploded, the first time this had ever happened. They had been using the Swift X-ray Telescope to observe light from a different supernova that had recently exploded in the same galaxy and been alerted by a five-minute blast of X-ray light that suddenly emerged from a different location, in one of the spiral arms. They decided to look at it more carefully with telescopes measuring other wavelengths, and just over an hour later saw the visible light flash of the supernova. The more energetic X-rays were emitted from the exploding star before the visible light, so they arrived first. Beyond the sheer excitement of seeing such an extraordinary event happening 'live', being able to see the supernova from its moment of explosion is helping astronomers better test their ideas about what is going on inside the star.

The nearest supernovae, the ones that happen in our own Milky Way, are truly striking. If a supernova were to explode in our Solar Neighbourhood it would briefly make the entire night sky look like daylight. One of the stars that we have our eyes most carefully on as a candidate for explosion is Betelgeuse, the orangey-red star at the top left shoulder of Orion that lies just beyond our Solar Neighbourhood. Betelgeuse is a former blue star that is less than 10 million years old but has already become an enormous red giant. We expect it to explode as a supernova 'any minute now', which in astronomical terms means some time in the next 100,000 years or so. In fact there is a slim chance that Betelgeuse has already exploded, as it is about 600 light-years away from us. That means that if anything had happened to it in the past 600 years, we would not yet know about it. When it does explode, Betelgeuse will suddenly look nearly as bright as the Moon in the sky.

We have many human records of supernovae happening in our Milky Way, even before the nature of them was understood. To early astronomers, they were simply unexplained brightenings in the sky. The earliest supernova we have on record was seen in the year 185, recorded by Chinese astronomers as a 'guest star' in the *Book of Later Han*. It took about eight months to fade, and there is even now a leftover remnant of the explosion seen as X-rays in the sky by the European Space Agency's XMM-Newton Observatory and NASA's Chandra X-ray Observatory, and as infrared light by NASA's Spitzer Space Telescope. At least ten more Milky Way supernovae have been recorded. The brightest was seen in April 1006, recorded by astronomers in Asia, the Middle East and Europe. A rock carving in Arizona, found in 2006 by the astronomer John Barentine, may also be a record of the star explosion seen by Native Americans. This supernova was relatively close by, 'only' 7,000 light-years from Earth. It was a quarter of the Moon's brightness in the sky, and visible in the daytime. The next Milky Way supernova was seen soon after, in July 1054. It was dimmer but still as bright as Venus and was visible for two years. It left behind the Crab Nebula, one of the most intensely studied astronomical objects, so bright that it can be seen with binoculars.

In November 1572 another Milky Way supernova was seen, this time watched by Danish astronomer Tycho Brahe.

This event had an important impact on astronomy, because Brahe realized that it had to be very far away, much further than the Moon, as it did not shift against the background stars as the Earth went round the Sun. At that time Europeans still believed that the heavens were fixed and unchanging beyond the planets, following Aristotle's concept of the cosmos, so this apparent sudden change of a 'heavenly' object was quite unexpected, and started to challenge the ideas of the time. This was in the era after Copernicus had put forward his Sun-centred model, but before it had been widely accepted. Brahe described his and others' observations in *Concerning the Star, new and never before seen in the life or memory of anyone*, published in 1573, and the event became known as Tycho's Supernova. Recent observations of the remnant left behind tell us that the star exploded about 8,000 light-years away.

The most recent supernova seen in the Milky Way came soon after, in 1604. Initially growing rapidly in size and so bright that it was visible in the daytime for three weeks, it was studied carefully for a year by Johannes Kepler, who wrote a book describing his observations; this one would become known as Kepler's Supernova. Galileo also observed the supernova and described his observations to a packed lecture hall in a series of public lectures at the University of Padua. Like Brahe he used the absence of parallax to demonstrate that it had to be far beyond the Moon, further proof that the distant stars were not unchanging, and further reason for Galileo and Kepler to support Copernicus' new model for the cosmos.

The nearest supernova in recent times was seen in February 1987, when a star was seen to explode in our neighbouring dwarf galaxy the Large Magellanic Cloud. It was visible with the naked eye, but here now was a wonderful chance to study a nearby supernova with modern telescopes. It was observed in great detail, starting just hours after the explosion, using a whole set of telescopes that could measure every type of light. That lucky event opened up astronomers' understanding of these dramatic explosions, allowing them to detect the heavier elements that it generated and to see the ejected matter being blown out, appearing as bright rings around the site where the star used to be.

We can see supernovae in galaxies beyond our own, or beyond the Magellanic Clouds, only with telescopes. The first substantial number of supernovae were found by Fritz Zwicky, a brilliant but notoriously combative Swiss astronomer who worked at the California Institute of Technology from the 1920s. He was to make many important discoveries, and with his colleague Walter Baade was the first to come up with the idea in 1934 that supernovae happened when normal stars turned into much denser stars. It was Zwicky who coined the name supernova. To find them, Zwicky pioneered the development of an 18-inch Schmidt telescope at California's Palomar Observatory, designed to be able to image large parts of the sky at once, making it more likely that these rare objects be found. He ended up finding more than 100 supernovae during his career.

There is now an impressive international network in place to watch for supernovae, with alerts set up to turn our largest telescopes to a new supernova as quickly as it is noticed, to catch its light before it fades. Astronomers get alerts in the middle of the night and have to make snap decisions about whether to redirect telescopes at these exciting events. It can be a difficult decision, as each hour and night of access to a large telescope is incredibly valuable and is not re-assigned lightly. But within weeks the main event will be over, so catching them in good time is crucial.

The objects left behind after a star explodes are among the most exotic in our whole universe. The white stars that started out just eight to twenty times heavier than our Sun will leave behind a neutron star. This is the strangest and densest of all stars. A white dwarf is already pretty extreme, an object as heavy as our Sun compressed into the size of our Earth. A neutron star is even more extreme. It is a couple of times heavier than the Sun, the rest of the original mass of the star having been ejected, but squeezed into a sphere several miles across. Neutron stars are so incredibly dense that a teaspoon of neutron-star matter would fall straight through to the Earth's core. The gravity on a neutron star is so strong that to escape its pull you would need to leave at almost half the speed of light, an impossible feat.

There are perhaps 100 million neutron stars in our Milky Way galaxy alone. Immediately after a supernova explodes, the leftover neutron star is as hot as a billion degrees and is thought to be spinning fast, rotating all the way round in less than a minute. This speed comes from the extreme shrinking in size of the star, similar to the way a figure skater spins faster when they draw their arms in close to their bodies. The stars quickly fade and cool to a mere million degrees. Neutron stars generally slow their spin too, but sometimes their gravity draws in gas from nearby stars which can spin them as fast as a few hundred times per second. As they spin, some of them send out jets of radio waves and X-rays, which we can see with specialized telescopes. We still do not understand exactly what goes on inside them: these are objects where the physical conditions are far more extreme than we can recreate here on Earth.

It was Walter Baade and Fritz Zwicky who suggested in a second 1934 paper that a star made of neutrons might exist, shortly after the neutron particle itself was first discovered. The stars were initially thought to be too faint to ever see, until Franco Pacini at the University of Florence worked out in 1967 that their spinning motion could produce a pulsating beam of radio wave light. Neutron stars have strong magnetic fields (a stronger version of the Earth's field that causes the Aurora Borealis), and as the magnetic field spins around the effect is to accelerate protons and electrons on the star's surface to form two jets of radio light. The beams emerge from the magnetic north and south poles of a neutron star, but magnetic north rarely aligns with the geographic North Pole around which it spins. Similar to a lighthouse, the beam of light that comes from those aligned appropriately will then sweep past our direction once every time the star spins round, so we see it as a pulsating signal.

By chance, in that very same year as the prediction was made, Northern Irish astronomer Jocelyn Bell Burnell discovered mysterious radio signals coming from a distant star, using a radio telescope that she built as a graduate student with her adviser Antony Hewish at the University of Cambridge. The pulses were spaced about a second apart, and they were so regular that for a while there was serious speculation that the signal might be being sent from a life form far away. The star's first nickname was LGM-1 for 'Little Green Men'. Soon, though, it was realized that this object, and others like it, were indeed spinning neutron stars emitting jets of radio light, and the objects quickly got their own name, the pulsar. Their discovery led to a Nobel Prize in 1974 for Hewish, but Bell Burnell was glaringly omitted from the honours.

For the heaviest stars of all, the blue ones more than about twenty to thirty times heavier than our Sun, a neutron star is likely not the final stage of stellar life. A neutron star can only manage to support itself if all that is left, after the supernova blows off the outer parts of the star, is a couple of times the mass of the Sun or less. The core of a heavy blue star is more massive than that, even after its supernova event, and the gravitational compression at the end is so intense that no outward-pushing force can balance it. Instead of a star, what comes into being is a black hole. Black holes are truly mysterious beasts. They form when you try to squash something at least twice the mass of the Sun into an area of space just miles across. The pull of gravity becomes so strong that the mass gets squeezed further into an extremely small space, perhaps infinitely small. You would have to leap out of a black hole faster than the speed of light to get away. Since nothing can travel faster than light, nothing can escape. Once in, there is no getting out. Indeed, the defining feature of a black hole is something that light cannot escape.

We can only speculate about what happens inside black holes, and right at their centre even our laws of physics break down. Strange things happen when gravity gets that strong. If you somehow fell feet-first towards a black hole, the pull of gravity on your feet would be so much stronger than on your head that you would be stretched out like a piece of spaghetti. Time would pass strangely too. Albert Einstein's theory of gravity tell us that time actually passes more slowly if you are near something very massive. For example, the rocks at the centre of our Earth really are a bit younger than rocks nearer the surface, where the gravity is not as strong. Odd but seemingly quite true. So if you imagined taking two twins and somehow sending one of them to spend time near a black hole, he or she would actually age more slowly than their Earth-dwelling partner.

We cannot see black holes directly, although astronomers can observe light coming from gas and dust that gets drawn in from companion stars to orbit around black holes. This disc of material gets very hot and sends out X-ray light, a sign that there is a black hole hiding within. But there is also another completely different way to see black holes. Their pull of gravity is so strong that it distorts space-time around them, in particular if two black holes orbit around each other. When two black holes circle around each other they will drag and stretch space around them, and we can actually search for their effect on space directly.

This behaviour is predicted by Einstein's theory of gravity, known as general relativity and published in 1915, which explains how space-time itself behaves. We will encounter this theory a number of times throughout this book. If we think about space a little like a rubber sheet, he showed that anything with mass deforms space, and something more massive deforms it more, just as a lead ball placed on the rubber sheet will bend it more than a foam ball. Objects, and light, take a path that follows the contours of the deformed space. Quickly pushing, or accelerating, a massive object through space will briefly change how much space is deformed, and this sets off a ripple known as a gravitational wave. This is what happens when two black holes orbit each other so quickly. They deform space-time itself, setting off a gravitational wave. This is a wave, but not a wave of light. It is really a stretching and shrinking of space itself. If such a wave passed through us, we would get momentarily taller and thinner, and then just afterwards a bit shorter and fatter, then back to taller and thinner, over and over again. It would happen to us, and to the whole Earth, and anything in its path. The effect would be real, but the change in our size and in Earth's size would be minuscule if we were just feeling the effects of a distant pair of black holes.

Until 2015 these waves had never been detected directly. We had never felt a gravitational wave come by. The year 2016 marked the triumphant conclusion to a fifty-year search for them, and the beginning of a new era of astronomy. The Laser Interferometer Gravitational-Wave Observatory, also known as LIGO, is an experiment that was thirty years in the making. Conceived in the 1980s, it now employs almost 1,000 scientists on three continents. It consists of a pair of gravitational wave detectors installed in Livingston, Louisiana, and in Hanford, Washington. Each one has two very long arms pointing at right angles to each other, each arm made of a tube a few miles long. As a wave reaches Earth and passes through, one of the arms will grow a bit, and the

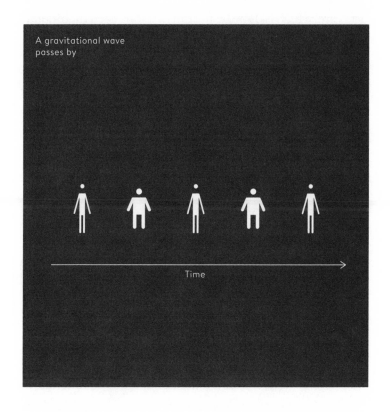

Figure 2.6
The exaggerated effect of a gravitational wave.

other arm will shrink a bit, and then vice versa, with the first arm shrinking and the other arm growing. This will keep repeating while the wave passes through the observatory. You can detect the wave passing by measuring how long the arms are, which is done by firing light along each tube, bouncing it off a mirror at the end and measuring how long it takes to get back. The idea is straightforward, but it requires exquisitely refined instruments, as the arms change length by very much less than a trillionth of a millimetre as the ripples go past.

The LIGO team joyfully announced their first detection of space-time ripples in February 2016, and the signal looked exactly as we would expect to see from two black holes spiralling and then colliding, turning more than three times the mass of our Sun into gravitational-wave energy. The collision happened about a billion light-years away from us, beyond our Milky Way, beyond our Local Group and local supercluster. The wave travelled for more than a billion years to reach us here on Earth on September 2015, passed right through, and continued its journey onwards. The LIGO team found the signature just days after turning on their upgraded experiment and announced a second detection from colliding black holes just a few months later. This was just the start of many new discoveries – three more spiralling black hole pairs sent ripples through the Earth in January, June and August 2017. The signal in August 2017 was also seen by LIGO's sister experiment, Virgo, an interferometer in Italy with a similar design to LIGO.

The LIGO experiment could identify roughly where in the sky the first signal came from, by seeing which of their two sites felt the signal first, but it could not locate the source with high precision. Immediately following the discovery of the signal, a well-choreographed follow-up took place. Astronomers round the world with access to telescopes that could look at the sky in all its different wavelengths, from radio waves to gamma rays, scanned the sky in the broad area to see if there was a burst of light alongside the burst of space-time waves. None was convincingly seen, and it is quite likely that these colliding black holes are visible only using gravitational waves.

Many of the stars in the universe are found orbiting each other in pairs. Neutron stars can themselves be part of a pair, orbiting around each other and spinning as they go, in an extreme cosmic dance. In 1974 the American astronomers Joseph Taylor and Russell Hulse discovered the first one of these using the 300-metre Arecibo radio dish. It was a pulsar orbiting around another neutron star every eight hours, with the pulsar itself spinning almost twenty times every second. By continuing to measure the stars over many years, astronomers observed a beautifully precise trend. The time taken for the two neutron stars to orbit each other decreased, which was exactly as Einstein's law of gravity predicted. The stars orbit each other so fast and with such strong gravity that, like the black holes, they deform space-time and produce gravitational waves. The stars use up some of their energy doing this, pulling them into a closer orbit. The star pair has now been observed for more than forty years and the same perfect trend continues, confirming the effect that Einstein had predicted in 1915. The two stars are expected to merge together in the end, but not for another few hundred million years. It was their discovery that inspired confidence in the physics and astronomy communities that gravitational waves would eventually be detected directly, and it won Hulse and Taylor the Nobel Prize in Physics in 1993.

On 17 August 2017 LIGO and Virgo identified another burst of gravitational waves passing through Earth, but this one was different. It looked like it had been created not from two black holes merging together, but out of the merging of two neutron stars, each a little heavier than the Sun. Unlike the pair found by Joseph Taylor and Russell Hulse, these ones were right at the end of their cosmic dance. The gravitational wave detectors felt the ripples in space-time that were produced in the last two minutes of their life, as these Sun-mass objects rapidly orbited 3,000 times around each other before colliding together. The wave reached the Virgo detector in Italy first and then passed on through the Earth to reach the LIGO detectors in Louisiana and then Washington, just milliseconds later.

Once again astronomers around the world mobilized in a choreographed campaign to point their telescopes at this rare and exciting event. Long-planned, they had been waiting for an event like this to happen. This time they struck gold. A signal from the collision appeared in every wavelength of light and was eventually seen by seventy observatories on all seven continents and from space. Two seconds after LIGO and Virgo felt it, a burst of gamma-ray light appeared in the sky and was seen by the Fermi and INTEGRAL satellites. These observations told astronomers roughly the direction in the sky but did not pinpoint the precise location. It was also still daytime in Hawaii and Chile, so astronomers using large telescopes had to wait for hours until nightfall, gearing up for the search.

Eleven hours after the initial detection, astronomer Ryan Foley and his team from the University of California in Santa Cruz were the first to see the intense infrared and optical light coming from this collision, and the first to identify the exact galaxy it had come from. This light had likely begun to be emitted right away when the stars merged, but it took those eleven hours to locate it in the sky. Foley had received the widely distributed text message alert and cycled straight to work, making a list of possible galaxies to observe with the Swope 1-metre telescope at Las Campanas Observatory in Chile, already in use as part of the Swope Supernova Survey. His team started remotely taking images of each candidate galaxy as soon as night fell in Chile. On the ninth image, post-doctoral researcher Charlie Kilpatrick announced he had 'found something'. Indeed he had. It was a galaxy 130 million light-years from Earth, where two neutron stars had ended their life in a phenomenal collision. Foley's team issued an alert out to the community, and other telescopes swung to take a closer look.

Numerous telescopes in Chile saw its optical and infra-red light, and as night fell in Hawaii astronomers saw it from there too. Fifteen hours after the initial event a signal in ultraviolet appeared. Nine days later the Chandra X-ray Observatory saw the X-rays. Last of all, sixteen days after the collision came the radio signal, seen by the Very Large Array in New Mexico. Delighted scientists presented their results and recounted their story of discovery at the National Science Foundation in the United States, and at the European

Southern Observatory's headquarters in Germany, on 16 October 2017. The journal paper had more than 4,000 authors from more than 900 institutions. It was a true international achievement involving the worldwide astronomical community.

The signal they were all seeing had come from a kilonova, the explosive result of the collision of the two neutron stars that left behind a single massive neutron star that likely collapsed into a black hole only milliseconds later. Predicted theoretically in 2010 by physicist Brian Metzger and collaborators, this sequence of light with all the different wavelengths had long been expected. The kilonova ejected material at a fifth the speed of light and produced huge amounts of gold and platinum: ten times the mass of Earth! Half of all the elements heavier than iron in the universe are thought to be made during these collisions of pairs of neutron stars.

Black holes, neutron stars and white dwarfs mark the end of the lives of the stars in our universe, but we have not yet looked carefully into their origins. How did our Sun come to be a giant ball of gas? We are pretty sure that it was born about 5 billion years ago, in a stellar nursery in our Milky Way galaxy. Its nursery would have been a giant cloud made of mostly hydrogen and helium gas. Gravity would have pulled that gas together into a cloudy clump inside our Galaxy. That cloud would have started out a cold place, more than 200 degrees below zero, and it was likely as big as our entire Solar Neighbourhood, tens of light-years across.

Inside that cloud, there would have been parts of it that

happened to be a bit more lumpy, a bit denser, than others. Those lumps would have gradually drawn more gas towards them until they collapsed into a swarm of swirling balls of hydrogen and helium. One of those balls of gas would have been our embryonic Sun. As gravity pulled it inwards ever more, it would have finally become hot enough for fusion to begin at its centre. We will return to this in chapter 5.

That was the point when our Sun as we know it was born. At around the same time, our neighbouring sibling stars were also ignited. Our Sun was light-years from its nearest star, but it was not entirely alone. Some of the nearby gas and dust would have been pulled into a swirling disc circling round the Sun, a bit like the rings of Saturn swirl around that giant planet. Within that disc, lumpier parts would have been drawn together, and one of those lumps ended up as our planet Earth.

If our Sun had been born before any massive stars had lived their lives, it would have been made entirely of hydrogen and helium. We think that those two elements were created right at the start of the universe, a topic we will explore further in chapter 5. Almost nothing else was around to begin with. That means that any planets circling that star would also have been made of just hydrogen and helium. Life as we know it would have been impossible on those planets.

The many elements that the planets, and we ourselves, are made of had to come from somewhere, and much of that took place at the very centre of older stars. It is there that we find the factories of the heavier elements, as the red giant stars fuse helium atoms together to make oxygen, carbon, nitrogen and other elements. And it is the supernovae, or the kilonovae from colliding neutron stars, that deliver all these elements out into space. The explosions violently force that mixture outwards, and it is from that mixture of gases that our Sun's nursery cloud was created. New stars are born from the remnants of old ones. Our Sun's cloud was still mostly just hydrogen and helium, but crucially for us it had those extra atoms that we need to make our rocky Earth and all the complex and diverse things that live on it.

We really are made of stars. The heavier atoms that Earth, and life on it, are made of were formed billions of years ago in a giant furnace right at the middle of stars that have now ended their lives. Of course our bodies grow and can make new molecules themselves, but they can only do that by replicating other cells, building on the elements that our Earth is made of.

We cannot see our own stellar nursery now, as our Sun is middle-aged and its nursery is long gone, but we can see many others scattered through the Milky Way. Other galaxies are full of them too. One of the most striking examples is the beautiful Horsehead nebula, which has majestic towers of gas shielding newborn stars. It was first recorded by Williamina Fleming at Harvard College Observatory in 1888. The dust and gas shrouding these stellar nurseries are very hard to see into with normal light, as the visible light from the baby stars gets absorbed by that gas and dust that surrounds them. With a visible-light telescope we would have no idea about what is going on within. Luckily we can now peer through the gas and dust by looking for infrared light coming from the young stars. Their infrared light can get through the cloudy shroud and can travel all the way to our telescopes, letting us see right into the places where the young stars are being born.

We have no reason to believe that our own Solar System, and the existence of our own planets in orbit around our Sun, are particularly unique. Our own Sun is a fairly typical star, and we think it formed in a normal way, as one of many stars forming from one of many gas clouds in one of many galaxies. Astronomers have therefore long thought that other stars likely host other solar systems, groups of other planets orbiting their own suns. Most astronomers would also guess that somewhere out there in our universe are other planets or moons that will have produced some form of life.

Until recently, though, we have had no way of knowing how many stars had their own planets, or what those planets were like. Planets do not send out their own light, or at least not much of it, so we cannot see them easily when we look out into the night sky, even with powerful telescopes. It is the stars that we see, and we are left to wonder what kinds of solar systems they are carrying around with them.

We are now living in a wonderfully transformative time in our quest to answer this question. Just in the last twenty years or so, astronomers have worked out how to find planets orbiting around distant stars in our Milky Way and have developed brand-new telescopes to look for them. What riches they have discovered! In 1992 the astronomers Aleksander Wolszczan and Dale Frail found two planets orbiting a pulsar, a surprising find at the time given the extreme events that tend to produce neutron stars. This discovery was quickly followed by the discovery of a planet orbiting a Sun-like star fifty light-years from Earth, found in 1995 by

Swiss astronomers Michel Mayor and Didier Queloz. Named 51 Pegasi b, where 51 Pegasi is the name of the star, it is a Jupiter-like planet that orbits around its star in just four days and is closer to its star than Mercury is to the Sun. A four-day orbit is far shorter than any planet in our Solar System, and was a hint of the incredible diversity of planets to come. Ten years later, about 200 planets had been found, and in 2018 this number had shot up to more than 3,000. NASA's Kepler satellite, launched in 2009, has had the largest impact on this haul. Discoveries of planets are now abounding, and the next decade promises to bring many thousands more with new space-based telescopes. The field of extra-solar planets, or exoplanets, is now a major new area of astronomy, and one that holds fascinating questions such as how unique our Earth is, and how unique life is on it.

We can find planets using a few different methods. It is hard to see a planet directly, as the light from its parent star is blindingly bright. The challenge has been likened to trying to take a picture of a firefly at night when it is next to a floodlight: the floodlight completely outshines the firefly. But a large planet, one quite far from its star, can be seen if you block out the star's light using a coronagraph, a shade that obscures the star in the same way that you could block the floodlight with your own hand. As of 2018 only about twenty planets have been found using this method, lying tens to thousands of times further from their star than Earth is from the Sun. Future space telescopes, including NASA's WFIRST satellite, aim to find many more.

Other methods rely on finding indirect evidence that the planet is there. The method used to find the first planets relies on the fact that a planet makes its star wobble. Since both the planet and the star have mass, the planet does not orbit around a point in the centre of the star. Instead the star and planet orbit around their common centre of mass, which lies in between the two, closer to the star, which is more massive. This jiggles the star around, making it move slightly towards and away from us each time the planet orbits around. We can test whether a star is moving in this way using the Doppler effect, something we have probably experienced with sound as a police or ambulance siren drives towards us or away from us. If something that sends out either sound or light moves towards us, we perceive its wavelength to be shorter than if it were not moving at all. The crests of the wave arrive more often. For sound, that makes the pitch higher than usual. For visible light, it squeezes the wavelength towards the shorter, 'blue' end of the rainbow. The opposite is also true, with a source of sound, or light, sounding lower in pitch, or looking redder in colour, than usual if it is moving away from us. The faster the source is moving, the more the shift in colour.

As the planet orbits the star, the starlight will appear first to have a slightly shorter wavelength as the star moves towards us, and then a longer wavelength as it moves away from us. The effect is subtle though, requiring a sophisticated spectroscope that can distinguish speeds of just tens of metres per second. The success of this method in finding 51 Pegasi b inspired a more focused search of other stars with large telescopes, resulting in the majority of the first couple of hundred planets detected in the late 1990s. The same wobbling effect will also cause a planet orbiting a pulsar to slightly disrupt the regularity of the pulsing signals. It was this method that Wolszczan and Frail used in 1992 to find the first extra-solar planets.

An increasingly popular method looks for planets crossing in front of their parent star as they orbit. During their transit, they block out some of the light from the star, just as if the firefly passed directly in front of the floodlight. It is a tiny dimming effect as the stars are so bright and the planets relatively small, but NASA's Kepler satellite was designed to look for exactly these transits. Most of the planets it has found are the rapidly orbiting ones, as a planet is easier to find if it passes in front of the star a number of times during the period of observations, which can last several years. A planet like Jupiter takes twelve years to orbit so would have been almost impossible to find using that method. This method can also only find the tiny fraction of planets whose orbits are aligned such that they fall in our sightline to the star.

Based on the census of planets made by Kepler, astronomers estimate that, in our Galaxy alone, at least half of the stars like our Sun have at least one planet larger than Earth that orbits in less than an Earth-year. It is quite likely that every Sun-like star has at least one planet orbiting around it, but those which take longer to orbit, or whose orbits are not aligned such that we can see them, have not been found yet. Some stars are already known to have entire solar systems with as many as eight planets.

The planets in our own Solar System were already a diverse bunch, but now the extended planet family has got vastly more varied. Many of them behave completely differently to

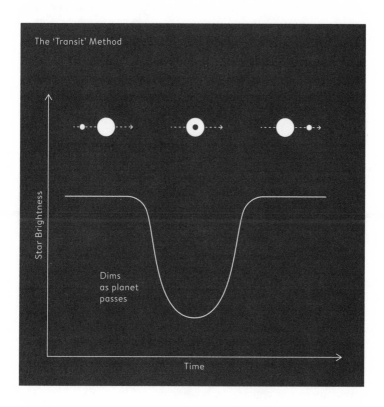

Figure 2.7
Finding a planet orbiting another star: it slightly dims the starlight as it passes.

our own. We have found huge gas planets like Jupiter that are much closer to their star than Mercury is to the Sun, that orbit their star in just hours or days. Some planets orbit around a star that itself is in orbit around a partner star, and a handful of planets have even been shown to orbit around both stars – these have been called 'Tattooine' planets, named after Luke Skywalker's fictional home. The most distantly orbiting planets have been found hundreds of times the distance of the Earth from the Sun and take thousands of years to orbit their star. The largest planets are many times more massive than Jupiter. Some rocky planets are so close to their star that their surface must be molten lava. Others are expected to be entirely covered in water.

Despite the richness of this new treasure trove of planets, we have only just scratched the surface of what is out there. For the time being, we have only been able to find planets orbiting stars in our own Milky Way galaxy, as galaxies beyond our own are so far away. We presume that there are just as many in other galaxies too. The set of solar systems that have been found are also by nature the ones most likely to be found using the methods we have available. Doubtless many more are waiting to be discovered.

One of the overarching goals of these planet searches is to work out how likely it is that a liveable planet like Earth could have formed around another star, and even better to find one. We define the habitable zone as the area around a star where planets could realistically have liquid water at their surface. This is known as the Goldilocks zone. Conditions must be just right for this: not too hot and not too cold, and with a solid surface. In our own Solar System,

Mars would be in that zone, as well as Earth, of course. A few dozen planets have now been found in the habitable zone of their stars, and we now think there are hundreds of millions of habitable-zone Earth-size planets in our Milky Way alone. The nearest found so far is just over ten light-years away, in our Solar Neighbourhood, and there is some evidence that there may be one around our very nearest star, Proxima Centauri. Of great recent excitement has been the discovery in 2015 of the Trappist-1 solar system, orbiting a star forty light-years away. It has seven planets, all comparable in size to Earth, and of which at least two are estimated to lie within the habitable zone. All orbit their star much closer than Mercury orbits the Sun, giving them years of between just two and nineteen Earth-days, and at least one planet has a liquid water ocean. This solar system, lying so close to ours and with such a richness of planets, will undoubtedly be the focus of significant future study.

It is not enough for a planet to sit in the habitable zone for it to be habitable. The conditions on the planet need to be right too, in particular those in the atmosphere. For example, an atmosphere heavy with sulphuric acid like Venus would likely be toxic to any life forms. We are now starting to look at the atmospheres of Jupiter-like planets and find out if they have the elements in them that we would expect to be necessary for life, and that might show hints of life already existing. There are a set of new satellites and telescopes setting out to advance this quest in the coming decades, including the European Extremely Large Telescope and the James Webb Space Telescope. There is now the exciting possibility that we might find an Earth twin in the foreseeable future.

In particular, we will want to look for elements like oxygen, ozone and methane in the atmosphere of a rocky exoplanet. Light from the host star, or chemical reactions with rocks on the surface, would usually diminish these elements, so seeing them would be a hint that life might already exist. It is the trees and algae on Earth that replenish the oxygen in our own atmosphere. The Trappist-1 system will certainly be part of this search.

As well as seeking to answer fundamental questions about the uniqueness of Earth as a host of life, the discovery of new solar systems is helping us to better understand how our own Solar System was created and how it might have changed during its life. The current idea that our own planets have migrated around inside our Solar System was partly inspired by seeing how planets were behaving around other stars. Broadening our horizon has let us see our own home in a brand-new light. And it's only been by broadening our own social horizons in astronomy – allowing women, and encouraging international movement and collaboration, for example – that we've made such astonishing leaps and bounds in the past century.

CHAPTER 3

Seeing the Invisible

Stars, to use the metaphor that they have given us, are the star players of the universe. On Earth, the Sun lights up our days, and other stars our nights. And if we panned out far enough to see the whole Milky Way, we would see a spiral disc full of stars, 100,000 light-years from side to side. Other objects pale in comparison. But, as we have seen, we are learning to detect a growing number of distant planets. And the universe contains a great deal else besides, some of it barely visible when viewed with telescopes sensitive to optical light, and some of it not visible at all.

Gas and dust, which we touched on earlier, are among the most important of these additional ingredients of the universe. Most of the gas in the Milky Way lies in clouds along the spiralling arms of the galaxy, where new stars are born. Much of that gas has been around since our Galaxy was formed billions of years ago, but some has been recycled from the remnants of older stars. As we learned in the previous chapter, these stellar-nursery gas clouds are extremely cold, more than 200 degrees below freezing. They produce small amounts of visible light, scattered from the stars within, but we can see them best through radio and infrared telescopes. There is also hotter gas throughout the Milky Way, some warmed up as hot as a million degrees by nearby stars. It is mainly hydrogen, but these atoms are so hot that they are broken up into their smaller parts, protons and electrons. This is known as 'ionizing' the gas, since an ion refers to a particle with electric charge, such as a hydrogen nucleus, which consists of a proton particle. The hot gas, which emits mostly X-ray light, spreads out further even than the Galaxy's stars, creating a faint halo of gas that envelops the Milky Way.

There are also tiny grains of cosmic dust spread throughout our Galaxy. This dust is quite different from the dust in our homes. A speck of dust on Earth can be as large as a tenth of a millimetre across or as small as around a thousandth of a millimetre. Most cosmic dust is smaller, sometimes just tens of atoms wide, more similar in size to particles found in smoke. Made of a mixture of elements, including carbon and silicon, we think that these dust grains were made in old, large stars and ejected out into space as the stars reached the end of their lives. The dust is everywhere in the Galaxy, though most, like the stars and gas, collects in the denser spiral arms. Mingling with the stellar-nursery gas, some of this dust will eventually form new planets when new stars are born.

Like gas, the tiny dust grains can be warmed up by the starlight that surrounds them, and this warmth generates infrared light. We now have beautiful images of neighbouring galaxies, including our next-door neighbour Andromeda, taken with infrared telescopes, including NASA's Spitzer Space Telescope. These give us quite a different view of Andromeda than standard pictures, showing up spiralling arms of dust where new stars form. We can take similar pictures of the Milky Way with infrared telescopes, but because we are taking pictures from within the Galaxy, we never get to see the whole thing at once.

The gas and dust are important parts of our Milky Way, but if we imagined all of the stars, gas, dust and planets as a bag of flour weighing a kilogram and containing about eight cups, we could take out all the gas with just one cup and all the dust in less than half a teaspoon. The planets, which number in the billions in our Galaxy, would account for just a pinch of flour. The rest of the bag would consist of stars.

There is also a giant lurking right at the middle of the Milky Way, even harder to detect than gas and dust. It is a huge black hole, a few million times heavier than our Sun – much bigger than the black holes left behind at the end of a big star's life that we met in chapter 2. Those might be only a few times or up to a hundred times heavier than our Sun. We do not yet know exactly how the one at the centre of our Galaxy got to be so much heavier than the rest. It might have started out as a 'normal' black hole that gradually ate up all the surrounding gas, growing ever heavier. Or it could have started out as a single enormous star that quickly collapsed into a black hole, or as multiple black holes that merged together.

We don't yet know its history, but we know it must be there because we can see stars flying around an invisible object at our Galaxy's core. The American astronomer Andrea Ghez, based at UCLA, leads a team using the 10-metre Keck telescope on Maunakea in Hawaii to study the centre of our Galaxy. They use a camera that can measure infrared light to see the stars through the dust that surrounds them. She and her team have been tracking the stars' paths around the black hole, known as Sagittarius A*, for more than two decades. Newton's laws of gravity tell us that a more massive object will cause things to orbit it faster. Ghez's team have worked out that only something with a strong pull of gravity, 4 million times the mass of the Sun and squeezed into an extremely small volume, could make the nearby stars orbit around it so quickly.

All the other galaxies in the universe are made of similar stuff to our own Milky Way, and many are thought to have large black holes at their cores. The balance of stars, gas and dust is different, though, depending on the type of galaxy, and the shape each galaxy takes is different too. We still do not fully understand how galaxies get their shapes and change over time, but back in 1936 Edwin Hubble divided galaxies up into three groups, described in his book *Realm of the Nebulae*. The first category is spiral galaxies. Like our Milky Way and our neighbour Andromeda, these are the most commonly observed. They are the rotating discs of stars with arms spiralling in to a larger bulge in the centre. In many of them the bulge is elongated into a denser bar of stars running down the middle.

Next there are the ellipticals. These galaxies are squashed or squeezed balls of stars, ranging in shape from spheres to elongated rugby balls or American footballs, or flattened M&M sweets. It is likely that many of these elliptical galaxies formed after the collision of existing galaxies, so they can end up growing many times larger than the spirals. We will probably end up in an elliptical galaxy ourselves when the Milky Way collides with Andromeda, many years in the future.

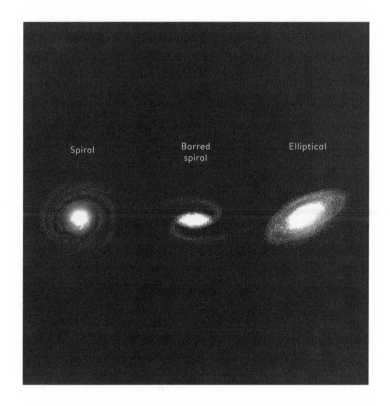

Figure 3.1
Categories of galaxies.

As a result of these collisions, the stars in the galaxies are often moving in arbitrary directions rather than spinning around the same way as they do in spiral galaxies. These elliptical galaxies also have much less gas and dust in them than the spirals, and so are barely forming new stars any more. Any of the more massive, short-lived white and blue stars are long gone in these galaxies, so they are dominated by the less massive, long-lived red stars. The final class of galaxies is the irregulars, which are neither spiral nor elliptical. Our small neighbours the Magellanic Clouds are of this type, with ill-defined, nebulous shapes.

The galaxies are not regularly spaced, as we learned in chapter 1: they gather together in groups and clusters, which are in turn gathered into superclusters. But if we were to imagine spreading all the galaxies uniformly throughout the universe, we would find them to be spaced a few million light-years apart. Remember, the size across the disc of stars might be 100,000 light-years, so the spaces between galaxies are on average between ten and 100 times larger than the size of each galaxy. These spaces are not all empty. The million-degree hot gas that envelops each galaxy also fills the space between the galaxies in a cluster, with typically a few protons and electrons in every shoe-box chunk of space. We cannot see this gas with optical telescopes, but it shows up shining brightly if we look with X-ray eyes. Astronomers do not yet fully understand how all of the gas got to be there, but there is a huge amount of it, so much that all the stars in all the galaxies make up only about 5 per cent of all the atoms in the observable universe.

*

Apart from black holes, all of these cosmic ingredients are things we can see in some way or other, using telescopes that pick up different wavelengths of light. There are plenty of bright things out in space to occupy our attention, but sooner or later we start wondering whether there might be other things filling our galaxies and our universe that we cannot see. An astronaut looking down on the Earth at night from space sees only the bright lights of the cities, the twinkling of human towns, villages, brightly lit houses. But even if they cannot see anything but lights, they know that there is so much more down there, the rich tapestry of our Earth covered in fields, valleys, mountains and oceans.

The lights can help our astronaut see the invisible, or at least to know the invisible. The brightest web of lights must be a city. The long line of lights snaking across the darkness must be a road. The bright, stretched-out ribbon of lights next to absolute darkness must be the coast, seaside towns marking the edge of the sea.

Like that astronaut looking down on us from above, we can know about the invisible in space by looking up at the lights that we can see. The key is gravity, because gravity does not care whether something produces light or not. Gravity only cares what something weighs. A heavy darkened object can pull a brightly lit thing, like a star, towards it. We see the star move towards some invisible point and we know that there must be something heavy there. We can imagine a similar thought experiment here on Earth too. I can imagine looking at a torch on a pitch-black night, with everything else in total darkness. Now if the person holding the torch lets go, we will see it fall quickly through the darkness and strike the ground. That happens because the gravity of the pitch-black Earth pulls the torch towards it. If the Earth were not there, the torch would just hang suspended in space. By looking at how the torch moves, I can work out precisely where the ground is. I could also figure out how heavy the Earth is by measuring how quickly the torch falls. If you dropped that same torch on the Moon, it would fall more slowly, because the Moon weighs less than the Earth.

Using the same idea, we can also weigh enormous objects far beyond Earth by seeing how fast smaller objects orbit around them. Let us imagine our torch again on the pitch-black night, but now we hurl it far up in the air. Imagine you can throw it so hard that it goes into orbit around Earth. Earth's gravity keeps pulling it inwards, but that initial hurl set it travelling in a circle, never to land back on Earth. If we also imagine that we have turned the Sun's light off for a moment, we would see our bright torchlight travelling around and around a pitch-black Earth. If we were watching this happen from afar, we would just see a bright light flying round and round in a circle. We would know there was something pulling it in that circle, even if we could not see it. The more the Earth weighs, the faster the bright light would move. Just by watching the light in orbit, we could figure out how heavy Earth is.

The same method can help us calculate the weight of vastly bigger objects, such as stars, galaxies and even clusters of galaxies. Without such a method, astronomers estimate the mass of cosmic objects by looking at them with a telescope and measuring their properties like colour and brightness. If it is a galaxy, they could measure how far away it was and how much light it was producing and, knowing the brightness of a typical star, could estimate how heavy it was. The more accurate method uses gravity directly, by calculating how fast things orbit around the object. It is a bit like the difference between estimating a person's weight from assessing how big they are and actually putting them on a weighing scale. Of course, the weighing scales give a more accurate answer.

Fritz Zwicky, the visionary Caltech astronomer of supernova and neutron star fame, realized in the 1930s that this second method could let him weigh an entire cluster of galaxies. It had only been ten years since Edwin Hubble had discovered that galaxies beyond our own Milky Way even existed, but Zwicky was quick to pick up on new ideas. He focused his efforts on studying the Coma galaxy cluster, which sits a few hundred million light-years away from our own Local Group of galaxies. In a cluster there is usually one or more larger galaxies at the middle, and other galaxies orbit around the cluster's centre of mass. Their motion is not confined to a flat disc, though, instead taking on a more spherical shape.

Zwicky studied how the galaxies were moving around within the Coma cluster, and they seemed to be travelling much more quickly than expected. He had already worked out how much all the galaxies in Coma should weigh by seeing how bright they were and working out how many stars they must have in them altogether. Based on that calculation, the galaxies were moving too fast. They were moving so fast, in fact, that it seemed as if Coma was many times heavier than it looked. The gas that fills the gaps in between galaxies could not account for all of the mass. In an attempt to explain this, Zwicky came up with the idea that there must be some sort of matter in Coma that he just could not see, that showed up in calculations only by using that second, direct-gravity method. He did not know what it could be, but in a Swiss journal article in 1933 he gave it a name: 'dunkle Materie', or 'dark matter'. This was a fascinating idea, a first clue that there might be more to space than meets the eye. It was to grow into one of the most elusive mysteries of our time, but Zwicky's discovery would lie dormant for decades, waiting for technology to catch up with him.

Zwicky noticed this puzzle of the excessive speeds when looking at an entire cluster of galaxies. In the years that followed, the same puzzle appeared to others studying individual galaxies. The heavier a galaxy is, the stronger the inwards pull of its gravity and the faster it will spin around, with all of its stars moving more quickly. The American astronomer Horace Babcock first noticed in the late 1930s that the stars in our galaxy neighbour Andromeda were moving up to twice as fast as expected. This was curious, but at the time Babcock did not connect this idea to any kind of missing matter. Twenty years later, in 1959, the Dutch astronomer Louise Volders measured stars in our other next-door-neighbour galaxy, Triangulum, and found a similar curiosity. It was also spinning too fast. Something was certainly strange, but at that point no one made a connection to Zwicky's idea about dark matter.

It was not until the late 1960s that things came together, and the idea that Zwicky had come up with so many years earlier took off in earnest. It had taken that long to get telescopes powerful enough to see far enough and in enough detail to measure the galaxies' motion properly. The credit goes to American astronomer Vera Rubin, who pioneered a set of transformational measurements of not just one or two but many tens of individual galaxies, working together with her colleague Kent Ford. Rubin was a trailblazer. She studied physics at Vassar College in the late 1940s and applied to do graduate work at Princeton University. It wasn't until 1961, though, that Princeton would admit its first full-time female graduate student. Rubin went to Cornell University instead to do a Masters degree, learning quantum mechanics from the famed physicist Richard Feynman. She then moved to Georgetown in Washington, DC, for her PhD, often having to attend lectures in the evening after looking after her two young children during the day. During that time she made a raft of important new discoveries, finding that galaxies were clustered together in space rather than being scattered randomly.

Rubin stayed at Georgetown after her PhD, continuing as a researcher and then professor and, in 1965, moved to the nearby Carnegie Institute for Science in Washington, DC. There she applied to use the 200-inch Hale telescope at the Palomar Observatory in California, the largest optical telescope of its time. It had been planned and developed in the 1920s by astronomer George Ellery Hale, a great pioneer in the building of world-class telescopes. He was also the person who had given Edwin Hubble a job working at the Mount Wilson Observatory, where he made so many important discoveries. Construction of Hale's Palomar telescope began in the late 1930s but it was only ready for use in 1949, years after Hale's death. Edwin Hubble would be the first to use it.

Until 1965, only men were allowed to use the telescopes at the Palomar Observatory, bringing to mind the days of the women 'Computers' banned from using telescopes at Harvard years earlier. But times were changing, and Vera Rubin swept these old-fashioned rules aside and became the first woman allowed to use the observatory's telescopes. The facilities were not yet designed with women in mind, and her former colleague Neta Bahcall, Professor of Astrophysical Sciences at Princeton, recalls how Rubin resorted to creating her own women's restroom by taping a paper skirt to the icon on the bathroom door.

Rubin worked with colleague Kent Ford, who had built a very sensitive spectrometer, the kind of instrument that splits the light from galaxies up into its different colours, or wavelengths. They installed it in the Hale telescope and used it to measure how quickly stars were orbiting around within a galaxy, by looking at the wavelength of light coming from different parts of the galaxy, and using the Doppler effect. If a galaxy is spinning round in a disc and we view it edge-on, the stars on one side will be moving towards us relative to the centre, appearing to have shorter wavelengths than usual, and the stars on the other side will be moving away from us, appearing to have longer wavelengths.

Rubin and Ford used this effect to work out how fast the stars were moving within each galaxy, and in particular how their speed changed further out from the centre of the galaxy. This was why they required such a large telescope, as they needed to be able to examine the component parts of

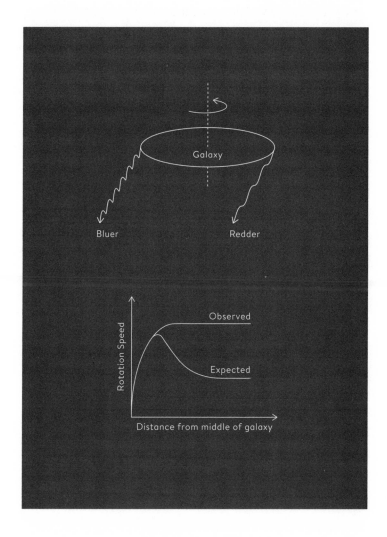

Figure 3.2
We can measure how fast galaxies spin using the Doppler effect.
Stars at the edges of spiral galaxies move faster than expected:
evidence for extra invisible matter surrounding the galaxy.

the galaxy in high definition rather than just see the whole thing as a smudge of light. They looked first at Andromeda, and then at more than fifty more distant spiral galaxies. In all of these galaxies they discovered an unexpected pattern. They found that nearly all the stars were orbiting around at about the same speed, regardless of how close to the middle of the galaxy they were. In our own Milky Way this would be about 140 miles per second.

They had expected to see something quite different. Based on what could be seen of the stars and gas and dust, the galaxies seemed to have a bright and dense bulge of stars and gas at their middle, surrounded by the less-dense disc of stars. It was expected that the stars beyond the edge of the brightest central bulge should go around more and more slowly as they got further from the very heaviest part of the galaxy, the part in the middle.

That was the idea anyway, but it did not fit Rubin and Ford's observations. The stars were still spinning round fast even at the very edge of the galaxy's disc, and the ones further out were not getting any slower. Something else had to be going on. Rubin realized that it all made sense if there was some extra mass in the galaxy that stretched way beyond the stars. The predicted motion matched what was seen only if each galaxy was really a few times larger across than it looked, with as much as 90 per cent of its mass completely invisible. It seemed as if the stars were just the bright lights of the city, embedded in the dark countryside of a much vaster galaxy.

This was strange and surprising, and the idea was initially met with great scepticism when they published their results in 1976. Perhaps it should not have been so surprising; Rubin herself recalled that this matched the idea that Zwicky had come up with forty years earlier. This was the 'dark matter', stuff that we cannot see but that does feel gravity. Now there were 100 galaxies telling the same story about dark matter that the Coma galaxy cluster had hinted at all those years before, and it soon became clear that Rubin and Ford had found convincing evidence that it existed. At around the same time, Jerry Ostriker and Jim Peebles at Princeton University used computer simulations to show that galaxies without any invisible dark matter could not stay in the disc formation that we see. They needed dark matter to keep them from breaking up. Everything started to make sense, and Zwicky's 'dunkle Materie' was alive again. What this stuff could be was a mystery: it is stuff that does not make any light, matter that we cannot see. It remains a mystery even today.

From what we can tell, there is dark matter in every galaxy, and in every group and cluster of galaxies. It also not only lies within and around those great cosmic objects but threads through space to form a great interconnected cosmic web. This web looks reminiscent of the neurons in our brains, blown up to gigantic proportions. It dominates over the visible stuff, with five times more mass in dark matter than regular atoms. The starlit parts of the galaxies are just the bright jewels in this larger dark network.

Our ability to understand and map dark matter is so much greater now than in Zwicky's day, not only because of our more advanced telescopes, but because of computers. The extraordinary leaps in computing that have taken place in the last few decades mean that computers are now almost as indispensable as telescopes, given how much more quickly than humans they can calculate. In helping us understand dark matter, computers allow us to build a virtual map of the universe, a model with computerized equivalents of all the dark matter, and in some cases the galaxies too, that make up the cosmic web.

To do this, we tell our computers what the laws of physics are, add instructions to simulate the presence of dark matter and ask the computer to work out what will happen to it all as time passes. We know, broadly, that gravity in the real universe will pull the dark matter into lumps, but our computers allow us to simulate the outcome in far greater detail than we could predict with pen and paper. We scatter dark matter throughout an artificial space, turn on gravity and then speed up the process in our computers. The computers work out what effect gravity would have on all the dark matter lumps, and over time we gradually see some lumps becoming larger and larger, with gaps being created in between. The simulations show in detail how the lumps of dark matter are linked together by these long filaments and sheets of dark matter threading through space.

To use a computer for this purpose, to work something out that no one has done before, involves writing some computer code, a set of instructions for what the computer should work out next. Just like with human instructions, an instruction to do something can be given in many different computer languages. Computer coding, coupled with telescopes, has become the foundation of modern astronomy, and our ability to understand our universe is closely

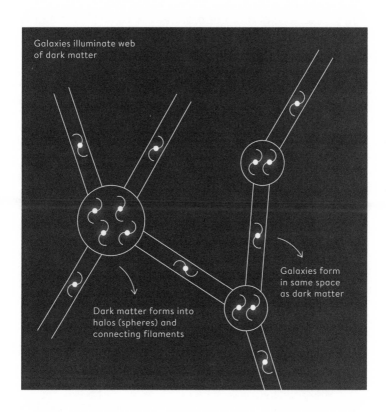

Figure 3.3
Cartoon of cosmic web of dark matter.

connected to how powerful our computers are. In the case of tracking how dark matter evolves, each sum is simple, as it just involves the simple law of gravity, but tracking how each bit of the dark matter interacts with all the other ones requires a huge number of calculations to be performed in quick succession.

The first computer simulations showing how dark matter forms into cosmic structures were pioneered in the 1980s by a group of astronomers known as the 'Gang of Four': George Efstathiou, Simon White, Carlos Frenk and Marc Davis. The simulations were attempting to explain some unexpected results seen from the first major attempt at a large survey of galaxies. Completed in 1981, the Harvard Center for Astrophysics survey, led by astronomer Marc Davis, mapped the positions of more than 2,000 galaxies, reaching to distances out beyond our Virgo supercluster. The astronomers saw the galaxies group together into nodes and long filaments, with large empty gaps in between. The galaxies should be found where there is the greatest density of dark matter, so this was the first indirect view of the dark matter web. At that time, the process by which the web formed over cosmic history was not well understood (a topic we will return to in chapter 5). It was therefore useful to be able to simulate what it might be expected to look like, and these first computer simulations did reproduce features that agreed with the observations.

Increasingly advanced surveys have been undertaken since the Harvard survey, recently including the Sloan Digital Sky Survey, which used a 2.5-metre optical telescope at the Apache Point Observatory in New Mexico to measure the positions of over 2 million galaxies. These surveys have refined our view of the positions of the galaxy groups and clusters in space. As our ability to measure the cosmic structures improves, so too must the fidelity of our simulations. Today, we can track tens of billions of dark matter chunks over billions of years in numerical simulations. They take a significant amount of computing power, with the state-of-the-art 'Illustris: The Next Generation' simulations produced in 2017 by a team led by German astronomer Volker Springel. These took thousands of computer-years to produce, achieved in practice by running thousands of computers, connected together within a supercomputer, for many months. These computer simulations tell us in much greater detail what the dark matter web of the universe likely looks like, and they also simulate the gas and stars in the universe, to help understand how the galaxies likely formed and evolved. The observations of galaxies are still, broadly, a good match to what these computer simulations predict.

Measuring how fast galaxies spin, or using galaxies as beacons, are both routes to finding the invisible matter. Einstein's theory of gravity showed us another stunning way to see it. Light will travel in a straight line through space unless it feels the pull of gravity from anything with mass. To understand this, it can help to start by picturing empty space a bit like a rubber sheet stretched out like a very bendy trampoline. A cosmic object like a galaxy sits on the rubber-trampoline and deforms it. The heavier the object is, the more it bends. A lead ball will make a bigger dip than one made of foam.

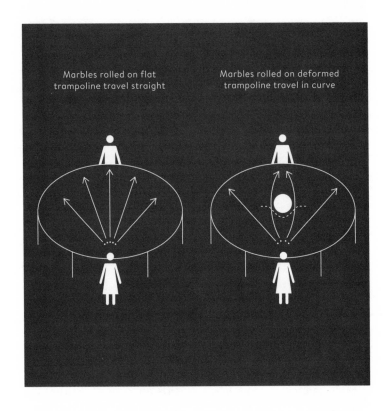

Figure 3.4
Picturing the path of light through space like marbles rolled on a trampoline.

We can now envision how a smaller object would travel through space by thinking about how it would move along the trampoline's surface. Let us imagine putting a very heavy ball right in the middle of the trampoline and then rolling marbles towards it. If we roll them straight towards the ball, they will fall right into the ball's dip and stop there. If instead we roll them far enough away from the ball to miss the dip completely, they will just keep rolling straight across the surface. Something more interesting happens if we roll the marbles just near enough to the ball that they reach the edge of the dip that it has made in the surface of the trampoline. They will skirt round the ball, changing direction slightly.

Now imagine a friend standing on the floor on the other side of the trampoline, directly opposite the ball. If you remove the ball and roll a marble across the surface of the trampoline just to one or other side of your friend's body, it will roll straight to that point. Now put the ball back on and roll the marble in the same direction. Instead of heading just to the side of your friend's body, the marble will swerve slightly when it encounters the dip of the ball and roll straight to the middle of your friend's body and into her hands. The marble would end up in the same place too if we rolled it at an equivalent spot on the other side of the ball, curving around in the other direction to reach our friend.

This is similar to how things travel in space, following the contours of the dips that are created by heavy objects. This was one of Einstein's great discoveries, to realize that mass and energy bend space, and that anything that moves must follow a path along a surface deformed by gravity. What is harder to imagine is how it works in three dimensions. The surface of the trampoline in our thought experiment started out as flat, two-dimensional, with the pressure of the ball creating a third, up-and-down direction. Space is like a three-dimensional version of the trampoline's surface, stretchy in all directions. A heavy object bends space just as a heavy ball bends a trampoline, but we cannot easily visualize this because our own three-dimensional brains simply cannot picture a bent fourth dimension. Regardless, we can imagine the consequences. Just like on the trampoline, we could roll a marble round the ball from the left or from the right towards our friend on the other side, but now we could also roll it over or under the ball.

Einstein realized that not only would small objects like meteorites or comets move through space like that, but rays of light would too. Imagine a bright source of light emitting rays towards a heavy object in space. The rays heading directly at the heavy object would simply hit it and stop in their tracks. The rays of light missing the object by some distance will just keep going straight. In between these rays, there will be some that come just close enough to the large object to enter the part of space that has deformed a little. They will curve around, some of them bending just enough to reach our eyes. They will have travelled right around the object to reach us.

This effect is known as 'gravitational lensing', with the heavy object acting as the lens that bends the light, and often a whole galaxy acting as the background source of light. The heavy object might be a cluster of galaxies, and if the bright galaxy is exactly behind the lensing object, we would see its light spread out into a ring of light centred on the lens. This

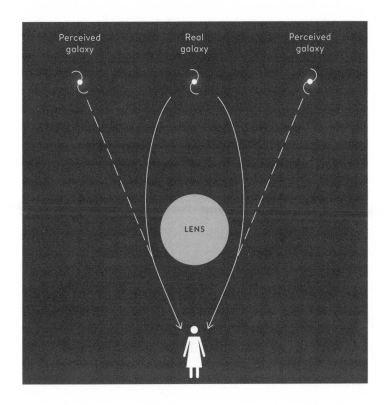

Figure 3.5
Light from a galaxy gets bent by a heavy object; the galaxy appears to be in more than one location in the sky.

is known as an 'Einstein Ring'. If the galaxy was a little offset, a little to the side, we would end up seeing a few copies of the galaxy. The light would behave like the marbles reaching our friend from both sides of the heavy object in the trampoline. The copies, or images, of the galaxy would be seen in the direction the marbles came from, and each image would be a little stretched out, like an arc of light instead of a bright dot.

Long before anyone observed this formation of multiple images of the same object, astronomers were eager to test the first main prediction of Einstein's theory, the simple fact that gravity could distort the path of light. Einstein's new theory predicted that light should curve twice as much as would be expected according to Isaac Newton's traditional theory. In 1913, Einstein wrote to George Ellery Hale in California, explaining his predictions and asking what conditions would be needed to measure the deflection of light from a single distant star being bent by the gravity of the Sun as the heavy lens. It would be a tiny bend, Einstein explained, less than a thousandth of a thumb-width held at arm's length. Einstein wanted to know if a total eclipse of the Sun needed to be happening to see this, when the Moon would block out all the sunlight, or whether a partial eclipse would work. Hale replied that a total eclipse was vital; otherwise any light from the Sun would completely drown out the light from the stars closest in position to the Sun.

Einstein had to wait for a total eclipse. The German astronomer Erwin Finlay-Freundlich hoped to make an expedition to Crimea in 1914 to observe such an eclipse and test Einstein's prediction, but the First World War intervened. It wasn't until 1919, one year after the war, that the deflection of light was finally measured, by Arthur Eddington and colleagues. Two groups set off to make the measurement, one to Sobral in Brazil and one to the tiny island of Príncipe off the west coast of Africa. They measured how much the light from the background stars was bent by the Sun, by comparing pictures of the stars' usual positions to their new positions when the Sun was bending the space in their path. They were slightly shifted, and the shift measured in Príncipe was consistent with Einstein's prediction. This result was joyfully announced and made front-page news in all the British national newspapers. It was Eddington's report of this expedition, at a public lecture on his return to Cambridge, that so inspired Cecilia Payne-Gaposchkin to become an astronomer.

It was only in 1936 that Einstein pointed out in a journal article that a natural consequence of the deflection of light was the existence of gravitational lenses that could produce multiple images of a single object. He had realized this years before but had not published the idea sooner, as he assumed that we would never be able to measure the effect. He had imagined that this would require light from a single distant star lensed around a nearer star, and the chance of getting two stars well enough aligned to see this happening is indeed extremely low.

Undeterred, Fritz Zwicky figured out in 1937 that, instead of single stars, entire galaxy clusters could be the heavy lenses that deform the space, bending the light coming from whole galaxies lying beyond. The alignment of galaxies would be much more likely to occur than the alignment of individual stars. Having made his observations of the Coma cluster in 1933, and finding hints of his dark matter, he would also have another way to weigh Coma to see whether there really was some missing matter. The more it bends light, the more massive it must be. It was a great scientific plan, but, as with many of his ideas, Zwicky was far ahead of his time. The telescopes of the 1930s were simply not good enough to measure the lensing of a galaxy. Here again, Zwicky had to wait for technology to catch up.

Finding a galaxy that has been gravitationally lensed can be done by finding multiple images of the same galaxy, located in the different directions the light took around the lens. It was not until 1979 that scientists managed to do this, when two images of the same quasar were seen by astronomers Dennis Walsh, Robert Carswell and Ray Weymann, using a 2-metre telescope at the Kitt Peak Observatory in Arizona. Quasars are among the brightest objects in the universe. First discovered in the 1950s, these are the bright cores of galaxies which have a massive black hole at their centre and a disc of gas drawn into orbit around it. A quasar forms when the disc of gas falls in towards the black hole and emits light of all wavelengths from radio waves to gamma rays. Quasars are thousands of times brighter than the Milky Way, and we can see them at great distances from Earth, further even than the Type Ia supernovae that we met in chapter 1. The light arriving from the most distant known quasar set off more than 13 billion years ago.

The astronomers spotted the lensing phenomenon in 1979 when they observed two quasars close to each other in the sky that looked identical, with the same amount of light being produced at every wavelength. They reasoned that they were likely two images of the same object, their light having been bent around a galaxy closer by. Named the Twin Quasars, the object is extremely far away from us, about 9 billion light-years away, while the galaxy doing the bending is 4 billion light-years away. They are out into the far reaches of our observable universe. At last, a gravitational lens had been found, but Zwicky himself never got to see it. He had died five years earlier.

One of the interesting things about seeing two images of an object is that you can actually be looking at images of the same object at different times, as the light travels different distances to reach us. That's pretty strange. It is as if we were to look across a room and see the same person in two different directions, one older than the other. We can get a sense of how this works by going back to the marbles on the trampoline. If we roll marbles at the same time from directly behind the large ball, opposite our friend, they will both arrive at the same time in our friend's hands, regardless of whether they go round the ball to the left or right. But if we shift sideways a bit, sitting slightly to one side of the large ball, then our friend will catch a marble rolled from that side sooner than one that has to take a longer, more curved, path around the other side, even if they set off together. The same thing happens with light travelling through space. Light from one of the Twin Quasars arrives on Earth more than a year before the other, so one of the images is the younger snapshot, one the older, of an identical object.

Just recently we had a spectacular example of this in real time, when astronomer Patrick Kelly at the University of California, Berkeley, used the Hubble Space Telescope in 2014 to

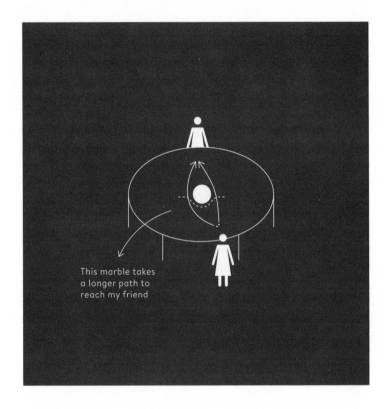

This marble takes
a longer path to
reach my friend

Figure 3.6
Two marbles can take paths of different lengths across the trampoline.
Light from objects in the sky can behave in a similar way.

study a distant galaxy cluster. He found a new supernova, naming it Refsdal after the Norwegian astrophysicist Sjur Refsdal. Its light was lensed by a heavy object in its path, so multiple images of the galaxy with the supernova appeared. By working out the shape and position of the heavy lens, Kelly and collaborators predicted that there should be two more images of the same supernova appearing at different times. One would have already appeared, but the other was predicted to show up a year later. Right as predicted, another image appeared in late 2015, exactly where it was expected. It was a beautiful demonstration not just of the effect of gravity on space, but also of our ability to predict and then test a scientific idea.

Today, we know of an abundance of cosmic lenses. A few years after the first discovery, a lens was discovered with four images of a quasar, and in 1988 astronomers detected the first Einstein ring, a smeared-out circle of light from a single galaxy. Now we have detected many thousands of gravitational lenses, with long arcs of light encircling massive cosmic objects. They are among the most beautiful and striking images we have of the cosmos.

What Zwicky had had in mind in the 1930s was to use this lensing effect to work out the mass of the galaxy or cluster acting as the lens. Today, astronomers are hoping to use gravitational lensing even more ambitiously to reveal the full extent of the dark matter in the universe. This should be possible because anything heavy, even if it does not produce light of its own, will bend the path of light. The more dark matter, the more it would make light bend. We can look at millions of distant galaxies, in clusters and superclusters beyond our own, and we can look at how their light has been bent by all the matter that it has passed on its way to our telescopes here on Earth. Most background galaxies are not particularly well aligned with a lens lying in their path, so the effect of lensing on most galaxies is a slight smearing of their shape rather than a creation of multiple images. We can use that shape-smearing to build up a three-dimensional map of the dark matter in most of the observable universe. We are just beginning to do this with current telescopes, and the next decade should see marvellous advances with the European Space Agency's Euclid satellite and the Large Synoptic Survey Telescope in Chile. We will return to some of these future projects in chapter 5.

We now know that the majority of the matter in the universe is invisible and find ourselves in the curious situation of having measured rather precisely how much of it there is, while at the same time having very little idea of *what* it is exactly. There is only one component of dark matter that we do know about. That is cosmic neutrinos, the tiniest particles we know to exist, each likely weighing less than a millionth of an electron, or less than a billionth of a hydrogen atom. They are everywhere, in phenomenal numbers. As many as tens of billions of neutrinos pass through your hand every second. We cannot see them, though, because neutrinos are essentially invisible. They do not send out any light of their own on any wavelength and they do not interact with the atoms in our bodies, or with atoms anywhere. Or at least, they almost never do. Many of them were created early on in the universe's life, and these are known as cosmic neutrinos.

Other neutrinos were created more recently in supernovae and in the Sun and other stars, and even in our own atmosphere.

We have known about neutrinos since Wolfgang Pauli came up with the idea in 1930. He was studying a particular kind of radioactive decay in which a neutron becomes a proton, or vice versa, inside an atom's nucleus. Known as beta decay, the process also produces an electron and a neutrino particle. Pauli thought up the neutrino as a way to make sense of the reaction, having only seen the electron and calculating that some energy from the atomic nucleus was missing. He didn't think this particle would ever be found, though, famously betting a case of champagne that neutrinos would never be detected.

Pauli started out calling this new particle a neutron, but in 1932 James Chadwick then took that same name for the heavier particle we now know. The Italian physicist Edoardo Amaldi came up with the neutrino's name later that year, to mean the 'little neutral one'. His collaborator Enrico Fermi started using the name at conferences, and it stuck. In 1933 Fermi submitted a paper to *Nature* explaining how the neutrino could be made in this process of beta decay, but his paper was rejected on the grounds of being too distant from reality. His model would later be proved correct. More than two decades later, in 1956, American physicists Clyde Cowan and Frederick Reines first detected neutrinos at the Savannah River Plant, an American nuclear reactor in South Carolina. They telegraphed Pauli with the good news, who soon sent them their case of champagne.

This first detection was of neutrinos being produced in

the nuclear reactor on Earth. But most of the neutrinos passing through us come from the Sun. They form as by-products
of the fusion at the Sun's core, while hydrogen is burning
to form helium. They fly out and away from the core of the
Sun and about eight minutes later reach Earth, taking just a
little longer than sunlight to reach us from the Sun's surface.
In the 1960s, American physicists Ray Davis and John Bahcall were the first to detect neutrinos arriving from the Sun,
using the Homestake experiment deep underground in the
Homestake Gold Mine in South Dakota. Bahcall calculated
the number of neutrinos the Sun should be producing, and
they soon noticed that only about a third of the expected
neutrinos seemed to be reaching Earth.

This mystery was soon designated the 'solar neutrino
problem'. Physicists and astronomers had figured out by
then that neutrinos come in three types, or flavours, known
as electron neutrinos, muon neutrinos and tau neutrinos.
In 1957 Italian physicist Bruno Pontecorvo had put forward
the idea that if neutrinos had mass, they could change their
type while flying through space. The solution to the problem, then, would be that the neutrinos were changing their
flavour, and that we had only been looking for and observing the electron neutrinos. There were many other ideas in
play though, and it took more than thirty years for scientists to show that this neutrino 'oscillation' was the right
explanation for the missing neutrinos. Between 1998 and
2002 the Super-Kamiokande experiment in Japan, and the
Sudbury Neutrino Observatory, in a nickel mine in Ontario,
both found that the flavours of neutrinos were indeed changing, using deep-underground experiments with thousands

of tons of heavy water to detect the neutrinos. This discovery won the leaders of these two experiments, Takaaki Kajita and Art McDonald, the Nobel Prize in Physics in 2015. And as a triumph for scientific prediction, the total number of neutrinos they found coming from the Sun ended up matching Bahcall's original prediction, made decades earlier.

Even after these complex experiments, we still don't know how much neutrinos weigh. Our best estimates suggest that the cosmic neutrinos make up between 0.5 and 2 per cent of all of the dark matter in the universe. In the early 1980s, many thought that dark matter might consist entirely of neutrinos, an idea inspired by the Russian physicist Yakov Zel'dovich. It soon became clear that this could not be true, however, as the web of dark matter would have come to look quite different. It is the gravity of the dark matter that pulls cosmic structures together, making the web in which the galaxies and clusters of galaxies sit. Neutrinos are so light that they fly through space close to the speed of light, and they try to escape from gravity's pull. This effect means that cosmic structures made out of neutrinos would be less clumped together than the ones made by slower-moving dark matter particles. By comparing the predictions of the Gang of Four's simulations, made of the slow-moving or 'cold' dark matter particles, with the grouping of galaxies seen in the Center for Astrophysics galaxy survey, astronomers concluded in the late 1980s that the cosmic web could not consist of purely neutrinos, otherwise the number of galaxy clusters and superclusters that we observe could not have formed, and that most of it had to be made up of heavier and slower-moving 'cold' dark matter particles. This finding has been confirmed again and again as teams of astronomers map out ever more galaxies and generate increasingly refined simulations.

Aside from neutrinos, the majority of dark matter is likely made of something that we have not yet encountered on Earth. A popular idea used to be that much of this cold dark matter might comprise familiar objects like stars, but which are too small to get fusion going at their cores, as well as planets and black holes. These would all be made of stuff that we do know of, but that generate almost no light. These objects are known as Massive Astrophysical Compact Halo Objects, or MACHOs for short. The problem, though, is that they would have to be big, from about the size of the Moon up to a hundred times the Sun. That's so big that we would have seen the effect of their gravity on stars. If they passed in front of a star, the pull of the MACHO's gravity would slightly bend the starlight, focusing it more towards us and making the star look a little brighter than usual. Astronomers have not seen enough of these occurrences for this theory for dark matter to work.

We then get left with another possibility, which is that the dark matter, aside from the neutrinos, could all be made up of a totally new particle, or set of particles. It has to be something that emits no light, or barely any, and that would travel through a person or a wall without stopping at all. This means it cannot be any of the atoms we know of, or the protons and neutrons and electrons that make up those atoms, or even the tinier quark particles that make up protons and neutrons. And, with some exceptions, it cannot weigh too little, like the neutrinos.

What we need is something new. One of the leading

possibilities is something called a Weakly Interacting Massive Particle, or a WIMP. The WIMP is a generic name that describes as-yet unknown particles that might be hundreds or thousands of times heavier than a hydrogen atom and that rarely interact with each other or with us. Like MACHO, the name WIMP is in common usage in physics and astronomy, and that the two names are semantically related is no coincidence. The WIMP was named first, and 'MACHO' was suggested as a joke by physicist Kim Griest in 1990, to emphasize the difference between these vast astronomical objects and the much smaller WIMPs.

Until recently, the WIMP thought most likely to be the elusive dark matter particle was a particle in the supersymmetry family. Supersymmetry is a physics theory that suggests that every particle we know of has a heavier partner. Each of the quark particles has a 'squark' super partner. The tiny electron has a heavier 'selectron' counterpart. It might sound fanciful, as none of these superpartners have ever been seen, but it is an elegant theory. The smallest of these hypothetical superparticles is called the neutralino, itself many times heavier than a hydrogen atom. The idea of the neutralino is particularly appealing to physicists, as it cannot break up into anything smaller so is not expected to interact with other particles. It is the end of the line.

There are other possibilities too in this family of supersymmetric particles, as well as other possible WIMP particles. Until lately, there were high hopes of creating these supersymmetric particles in the Large Hadron Collider at CERN, the European nuclear research facility near Geneva, and actually finding the dark matter particle. Despite the great success of the CERN experiments in finding another elusive object – the Higgs particle – there is no sign yet of supersymmetry. That doesn't mean it doesn't exist, but it means it is at least hiding out of sight, which has raised doubts in some people's minds about how likely it is to exist at all. Perhaps it just does not describe reality.

The Large Hadron Collider is not the only tool we can use to find these elusive weakly interacting particles of dark matter. We can also hope to actually catch one in a detector. There are a host of experiments underway in old mines deep underground to try to find them, including the Large Underground Xenon experiment in South Dakota. And, at the same time, we peer at the skies too to get hints, trying to see some sign of these particles interacting and sending out some light signal. Nothing has yet been seen.

Other interesting theories of what the dark matter might be include a tiny particle called an axion, or an extra heavier neutrino particle, or particles that only arise if we actually live in more dimensions than we know. Or perhaps we should not look for just one particle. It might consist of a whole family of dark particles. It also might be something that no one has ever yet imagined.

We are now in this odd situation where we think that the wider world we live in is mostly invisible, that there is some new 'stuff' throughout space that weighs in total five times as much as the atoms that we know of, and it is everywhere: here on Earth, in our Solar System, in the Milky Way and forming the cosmic skeleton in which the visible galaxies and clusters of galaxies are embedded. Almost all of our evidence that this stuff exists, of course, is based purely on looking at the effect that the dark matter's gravity has on visible things.

On that basis, it is common for astronomers to ask themselves if it is possible that we have misunderstood how gravity works in the first place. Is dark matter somehow an illusion? Could our laws of gravity, which tell us how things should move in the presence of other objects, need some refinement? It is an obvious question to ask, and an Israeli physicist, Mordehai Milgrom, did exactly that. In the early 1980s he came up with a theory called Modified Newtonian Dynamics, also known as MOND, as an alternative way to explain the rotation of galaxies seen by Vera Rubin and Kent Ford. His idea was that force decreased more slowly as the distance from a massive object increased, compared with in Newton's or Einstein's usual laws of physics. In fact, Rubin herself initially found this idea more attractive than the need for a new particle.

This modification to the laws of gravity would only kick in when gravity was extremely weak, so it would not have been noticed on Earth or for objects in the Solar System. Only in the outer reaches of a galaxy would it start to take effect, conveniently explaining the spinning galaxies. This theory has a number of significant weaknesses, though, as it cannot explain many other observations.

An example of this is a striking observation made in 2006 by American astronomer Douglas Clowe and collaborators, which gave extra credibility to ideas about dark matter. What they saw was the Bullet Cluster, two huge galaxy clusters in the aftermath of a collision. The astronomers were seeing an image of a cluster of galaxies that had shot straight through another, larger one, apparently at millions of miles per hour. What happens when you try to push one cluster of galaxies through another? Ignoring dark matter for now, we can think of the cluster as made of galaxies and very hot gas. The galaxies themselves are relatively small, and the ones in the smaller cluster would pass through the larger one without colliding. But the gas would get slowed down, interacting with the gas of the other cluster, and would lag behind. This was indeed seen to be the case by taking pictures of the Bullet Cluster with optical light and X-ray light. The optical light showed up the galaxies, and the X-ray light showed up the gas. The gas was clearly caught lagging behind the galaxies in the bulleting cluster.

Now what about the dark matter? Well, in a galaxy cluster without dark matter, most of the mass should be where the hot gas is, as this weighs much more than the stars. If the galaxy cluster does have dark matter, the dark matter should have shot right through the cluster along with the galaxies. Because it doesn't interact with anything, it would not have been slowed down like the gas and, in combination with the galaxies, would have formed a mass heavier than the gas. Conveniently, we can use gravitational lensing to find out where the heavy parts of distant objects are, and the team of astronomers looked at how this pair of clusters distorted the light of distant galaxies lying behind the clusters. By doing so they could calculate which location held more mass: that occupied by the gas or that occupied by the galaxies. They discovered that the space occupied by the galaxies held the most mass. It must therefore have contained something other than the galaxies: dark matter.

The discovery of the Bullet Cluster made a strong argument in favour of dark matter. Most astronomers and physicists do not think we are fooling ourselves about its existence. Dark matter really does seem to be part of our world, but until we have found it we try to keep our minds open, continuing to wonder what it might be.

CHAPTER 4

The Nature of Space

So far we have found out how the universe is put together. We have climbed up the cosmic ladder, increasing the scale step by step from solar systems, to stellar neighbourhoods, to galaxies, to groups or clusters of galaxies, and finally on the very largest scales to superclusters. We have also found out about the visible and invisible things that belong in these different realms of the universe, including planets, stars, black holes, clouds of gas where stars are born, cosmic dust, hot gas that permeates the galaxy clusters and the as-yet unidentified dark matter. We have focused on what the universe looks like now, touching only briefly on its past and how, the further into space we look, the further back we see.

In this chapter we will find out more about the nature of space itself. Is it infinitely big, and has it always existed? These questions will take us as far back in time and out into space as we can go, bringing us to the very beginning, the birth of our universe.

As we saw in chapter 1, it was only in the 1920s that people came to realize that our own Milky Way was not the sum total of the universe. The Great Debate in 1920 between Heber Curtis and Harlow Shapley concerned whether anything at all existed beyond our Galaxy, and in particular whether faint smudges of light seen in the night sky were within the Milky Way or far beyond. The debate was resolved by Edwin Hubble's observations of the pulsating Cepheid stars. Drawing on Henrietta Leavitt's work on patterns in the Cepheids' luminosity, he established that the stars in the smudges were too faint and therefore too distant to live in our Milky Way, identifying the nebulae as entirely new galaxies beyond our own.

This came just a few years after Albert Einstein had developed his theory of general relativity, the magnificent theory that describes how space itself behaves. As we learned in chapters 2 and 3, Einstein explained that matter bends space in the way that the large ball we described in the previous chapter makes an impression in a trampoline. The more massive the object, the more it bends space. The more deformed the space is, the more it will affect the paths of nearby objects. The mass of the Sun determines Earth's orbit in this way, just as on a grander scale the mass of galaxies determines how they pull each other around in a galaxy cluster.

Einstein's beautiful theory not only tells us how objects in space interact, but also something about how the whole of space behaves. His theory predicted that space should be continually changing. If you imagine scattering matter throughout a universe, it should not just make dimples in the space, but should also start shrinking space. The gravity of all the different objects should gradually make everything pull inwards towards everything else. The trouble with this prediction was that Einstein himself hated the idea: he was convinced that the universe was unchanging, that it must be the same now as it was in the past and would be for ever more.

At the time, the idea that everything was stable seemed to match the reality of what we could see in the sky. Before Hubble revealed the universe beyond the Milky Way, there was no evidence of any significant change occurring.

Einstein got around the problem of a changing universe by modifying his theory to add something he called the Cosmological Constant. It was energy wrapped up in empty space itself. This energy would try to make space grow and would exactly balance the opposite tendency of space to shrink, keeping things precisely still. This was a clunky fix, and it was certainly not the only possible way to explain the behaviour of the universe.

One of the first people who seriously entertained alternative ideas about the fact that space might be changing was the Russian physicist Alexander Friedmann, who studied physics at St Petersburg University, before the First World War and service in the Russian air force interrupted his academic pursuits. After the war, Friedmann carefully studied Einstein's new theory of general relativity, and in the early 1920s he realized that there was a simple explanation for how the whole of space should behave, assuming that things should look the same in all directions and from every vantage point. He used Einstein's equations to come up with the notion that the universe could actually be expanding rather than shrinking, and that it should certainly be changing.

Friedmann published his results in 1922 and 1924 in the German physics journal *Zeitschrift für Physik*. Einstein initially responded to his 1922 paper with great negativity, dismissing Friedmann's description of the expanding universe as impossible. Friedmann wrote to Einstein to explain the calculations that had led to his conclusion, but Einstein was busy travelling the world, and it was months before he finally learned of the letter. Once he did, he acknowledged that Friedmann's results were correct, that an expanding universe was theoretically possible, but he still didn't find the idea compelling. He remained convinced that the universe was static. Friedmann would soon be proved right, but he was to die tragically young of typhoid fever in 1925, never to know what an important contribution he had made to our ideas about the universe.

A few years later, the Belgian physicist Georges Lemaître independently used Einstein's equations to come up with a similar conclusion about the behaviour of space. Lemaître was not only a scientist but also a priest, a vocation he had chosen when he was only nine years old. To him these twin pursuits were equally important, and he studied physics and mathematics at the same time as his theological training, after serving in the Belgian army in the First World War. In the early 1920s he spent formative time working with Arthur Eddington in Cambridge and with Harlow Shapley at the Harvard College Observatory, completing a PhD in 1927 at the Massachusetts Institute of Technology. On his return to Belgium, Lemaître used Einstein's theory to work out that space had to either grow or shrink. Like Friedmann before him, he concluded that a static universe was not an option. He published his results in 1927 in an obscure Belgian science journal, the *Annales de la Société Scientifique de Bruxelles*, but it was written in French and read by very few people outside Belgium. He described his work to Einstein at the Solvay Conference in Brussels that same year, and Einstein

famously told him that 'your maths is correct but your physics is abominable'. Einstein was resolute in refusing to think
that a changing universe could describe reality.

To help us understand the ideas that shaped these debates,
and the measurements that were made to eventually resolve
them, we will now take a step back and consider what it actually means for space to grow or shrink, and how we might tell
if either were happening. It is hard even to define what space
is. We might, for example, think of it as the gaps in between
things. Near home that would include the gaps between the
Earth and the Sun, then the gaps among the Sun and the
nearby stars, and the gaps among the stars. On the grandest
scales it would be the gaps among the galaxies and galaxy
clusters that fill our universe. It is perhaps better, though, to
think of space as not just the gaps, but as everything: all the
objects in space as well as the gaps among them.

   To start to imagine a space that is growing or shrinking,
we will use an analogy that is imperfect but which helps us
predict how we might view such a space. Let us think of an
ant living in a one-dimensional universe, where the ant's
entire universe is a long piece of elastic, the sort you might
find in a waistband, stretched out in a long line. The ant is
confined to walking back and forth along that piece of elastic. She cannot move sideways, up, or down. We now make
that ant's universe grow by taking hold of each end of the
elastic and pulling it gently. As we pull, the elastic stretches
out a bit. Every part of it elongates a little. And even though
we are pulling it at each end, the stretch is not coming from
one single place in the elastic. It is coming from everywhere.

We then do the reverse, making the ant's universe shrink. We do this by gently releasing the tension in the stretched elastic. It gets shorter, shrinking everywhere. There is no single thing at the middle of that piece of elastic that is pulling it inwards. The inward pull happens all along the line. This is somewhat similar to how space is behaving. When it grows, it grows everywhere, like the elastic. When it shrinks, it shrinks everywhere.

There are some obvious differences between this model and reality, beyond the fact that we live in three dimensions. One is that nothing in reality corresponds to our holding on to the ends of the elastic, or even to the idea that the elastic has ends at all. We do not think that space has any ends. The two ways that this could be achieved in the one-dimensional elastic space would be either for that piece of elastic to stretch on and on in an infinitely long line, or for each end to be connected together, forming a loop. The infinitely long line is hard to imagine, but for our thought experiment we just have to envision the piece of elastic on which the ant stands. As it stretches, we can watch it grow.

Now, how could our ant figure out if her universe was growing? How could she see this from within her universe? She would need some markers along the elastic to identify and measure. Let's assume, then, that we have stapled round stickers along the elastic. Now imagine stretching out the elastic again, pulling it from each end. As we pull, we would see all the stickers move apart from each other. If they start spaced out an inch apart, they might end up two inches apart.

That is what we see when we are holding the elastic, looking on from above. What does the ant see from her vantage

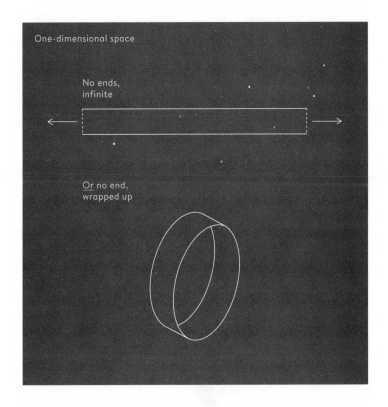

One-dimensional space

No ends,
infinite

Or no end,
wrapped up

Figure 4.1
Two different ways for a one-dimensional space to have no ends.

point? Let us put our ant on one of the stickers, looking either straight ahead or backwards down the line. As we stretch the elastic, our ant will see all the stickers appear to move away from her. The nearest sticker to her will appear to move from an inch away to two inches away. The next one over will appear to move from two inches away to four inches away. The distances from our ant to all the other markers will appear to have doubled. She will see exactly the same thing whether she looks forwards or backwards down the elastic.

To our ant, it will look as if the more distant markers actually moved away faster during the stretch than the closer ones, as if they travelled a greater distance from her during the time taken to stretch the elastic. The more distant a sticker, the further and faster it appears to move. If it appears to have moved twice the distance, then it appears to have moved at twice the speed. As the markers all appear to move away, they follow a precise pattern, with the more distant markers appearing to recede faster.

This pattern would be true whichever marker we moved our ant to. She would see the same thing. She would perceive herself to be at the centre of everything, with all the markers in the universe moving away from her. In reality, this would just be her point of view. It would be the natural consequence of living somewhere in a space that is growing everywhere. If instead the ant were living in a universe that was not growing, then there would be no particular pattern to how the markers around her would appear to move. On average they would not be moving at all.

This thought experiment with the elastic is one way to help us picture the effects of a real expanding universe. The

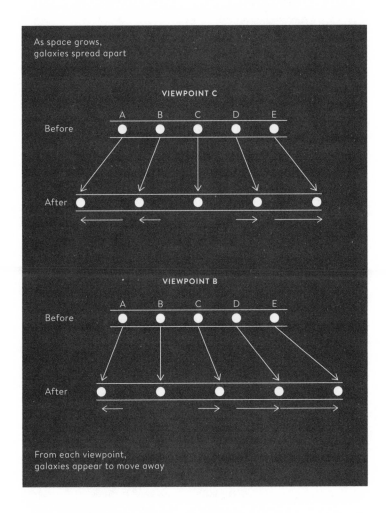

As space grows,
galaxies spread apart

VIEWPOINT C

Before   A   B   C   D   E

After

From each viewpoint,
galaxies appear to move away

VIEWPOINT B

Before   A   B   C   D   E

After

Figure 4.2
How an expanding one-dimensional space looks from different
viewpoints.

spaces among cosmic objects get larger, like the spaces between markers on the elastic, but of course our universe is not one-dimensional. If our ant were to live in two dimensions we could think about her living on a stretchy rubber sheet covered in sticker-markers. As with the elastic, we can imagine that sheet stretching. As it grows, it stretches everywhere. The stretching has no centre, and as the sheet grows all of the markers move farther apart from each other. If we again imagine an ant living on one of them, she would now see markers appear to move away from her in all directions. Just as with the elastic, the more distant markers would appear to recede more quickly. An ant living on any one of those markers would see the same thing.

Now imagine space growing like a stretchy elastic in all three dimensions. For a physical analogy, we might think of a bread dough, scattered with raisins. The yeast inside the mixture causes the dough to rise, to expand in every direction. It does not grow from any one particular place, and seen from afar, all the raisins move further apart from each other as the dough expands. Seen from the viewpoint of a particular raisin, all the raisins around it will appear to move away as the dough rises, and the more distant raisins will seem to recede faster.

These analogies of elastic, a rubber sheet or dough may help us imagine what a growing space might look like, but in each case the analogy breaks down when we think about the edges or ends of the material. Space does not have edges. There are two broad ways of getting rid of the edges. Either space just continues on and on infinitely far, with ever more elastic or ever more dough. Or, it loops round just like the piece of elastic, but now in three dimensions. This is a

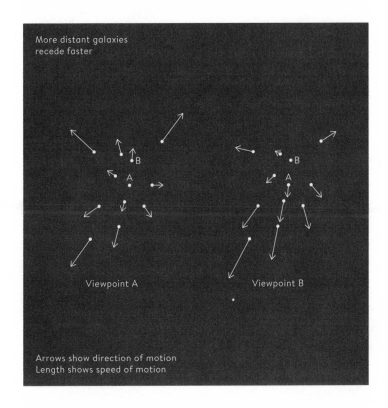

Figure 4.3
How an expanding two-dimensional space looks from different viewpoints.

difficult concept to visualize, and we will return to what that means later in the chapter.

What about the markers in space? The closest thing to the stickers, or the raisins, are the galaxies that fill space. They act like our cosmic markers. We can sit here in the Milky Way and look out at galaxies beyond our own, to see what pattern of movement they follow. As he wrote in his paper of 1927, Georges Lemaître worked out that we could test whether our universe was growing by looking for the sort of pattern we described in the analogy of the raisins in the dough. The galaxies should all appear to be moving away from us, and the more distant ones should appear to be receding faster. Within a gently expanding space, the galaxies themselves would not be expanding, since the gravity that holds them together is stronger than the stretching of space.

Lemaître showed that this expanding space model was compatible with new measurements of distant galaxies. How did he do this? Finding a set of distant galaxies, and calculating whether they are all moving away from us, sounds fairly straightforward, but the distance and motion of galaxies are technically very hard to measure. Let us start with the distances. At the time, the best distance estimates used the brightnesses of galaxies. Edwin Hubble compiled a set of these distances in his 1926 paper 'Extragalactic Nebulae', based on 400 nebulae, as they were known at the time, that had already been observed. Hubble assumed that all these distant nebulae had the same intrinsic brightness, so their distances could be estimated by measuring how bright they appeared from Earth. More distant nebulae appeared fainter.

Next, to work out how fast a galaxy is moving away from

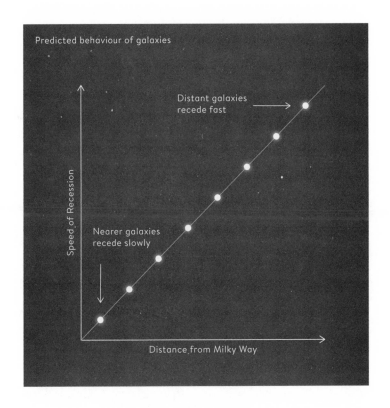

Figure 4.4

In an expanding space, the more distant galaxies should appear to recede faster from the Milky Way.

us, we cannot simply measure its distance from us at some time, and then its new distance again a little later. The relative distance it will have moved during that time would just be too tiny. Instead, we rely on something easier to measure: the colour of the light coming from a galaxy.

As Vera Rubin demonstrated with the spinning galaxies, light appears 'redder' than usual, or with a longer wavelength, when sent out by something moving away from us. Something moving towards us looks 'bluer' than usual, having a shorter wavelength. The same thing applies to entire galaxies. A galaxy will look redder than usual if it is moving away from us, or even if it just appears to move away from us because space is growing. If we can measure the colour of light coming from a galaxy, the only challenge then is knowing what its usual colour would have been if it weren't moving. Once we know both these wavelengths, we can work out the apparent speed of its motion.

Here we turn back to astronomers' earlier discoveries of what stars are made of. As we know, stars are made of mostly hydrogen and helium gas, but contain traces of other elements too. The different elements emit and absorb light at very specific wavelengths. In chapter 2 we encountered the idea that if we measure the spectrum of a star, we find dark lines where the star's atmosphere has absorbed a particular wavelength of light. The same goes for a galaxy full of stars. If we measure the spectrum of a galaxy, it too will have a pattern of dark absorption lines, and bright emission lines, at particular wavelengths. The faster a galaxy is moving away from us, the more we will see the lines shift to the redder end of the spectrum.

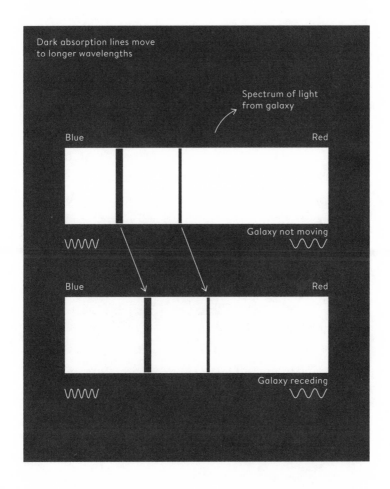

Figure 4.5

If a galaxy is receding, its light will be received with a lengthened wavelength, known as redshift.

This shift in the wavelength of light is called the 'red-shift'. Even before knowing exactly what stars were made of or even that there were other galaxies beyond our own, the spectra could still be used to measure this shift. Vesto Slipher, the astronomer at Lowell Observatory in Flagstaff, Arizona, who later persuaded Clyde Tombaugh to go out looking for Pluto, made vital observations of the spectra of a set of galaxies, including the first-ever observation of a spectral shift in 1912. This was the spectrum of what turned out to be our neighbouring galaxy Andromeda, which at the time was not known to be outside the Milky Way.

Slipher then measured fourteen other nebulae in all directions around us on the sky. He found Andromeda's spectrum to be shifted towards the bluer end, implying that it was moving towards us, and he measured its speed to be 300 kilometres per second, much faster than any object within the Milky Way. He found the spectra of almost all of the others to be shifted towards the redder end. The majority of the nebulae appeared to be moving away from us. At that point, in 1915, it was still not known that these objects were outside the Milky Way, although their high speeds gave a clue that they might be. It was ten years later, in 1925, that Hubble showed convincingly that the nebulae were indeed spiral galaxies far beyond our Milky Way.

Georges Lemaître realized the importance of Slipher's findings. Recall that if galaxies are moving in the same pattern as the raisins in our hypothetical dough, then the more distant the galaxy is from our fixed point, the faster it will appear to be moving away from us and the redder its light will be. If more distant galaxies indeed have spectra whose lines are shifted to longer wavelengths than closer galaxies, then we have found evidence for the expansion of space.

Lemaître measured the pattern, finding that more-distant galaxies appeared to be moving faster. The galaxies around us really did appear, on average, to be moving away from us, as you would expect in an expanding space. The local motion of Andromeda towards us could be explained by the pull of gravity within our Local Group. Lemaître used Slipher's redshift measurements and Hubble's distances to estimate the rate of expansion of the universe, finding a rate of just over 600 kilometres per second per megaparsec. These units mean that if two galaxies were separated by a distance of a megaparsec, which is equal to a million parsecs or just over 3 million light-years, the expansion of space would cause them to be receding from each other at 600 kilometres per second. Galaxies separated by twice that distance would be moving away from each other at twice that speed.

So Lemaître showed in 1927 that space was likely expanding. But very few people noticed his result in the Belgian journal. At that time Edwin Hubble had started a programme to make more accurate measurements of the distances to Slipher's galaxies, working with a talented assistant, Milton Humason, at Mount Wilson Observatory in California. By measuring Cepheid variable stars, and using Leavitt's Law to relate their pulsation rate to their intrinsic brightness, Hubble and Humason measured the distances to twenty-four of Slipher's galaxies, no longer relying on the assumption that they all had the same intrinsic brightness. Although it was still difficult to measure precisely, by now the pattern was unmistakeable. Most of the galaxies did appear to be moving away from us, and the more distant ones were receding faster.

Hubble published his findings in 1929, in 'A Relation between Distance and Radial Velocity among Extra-Galactic Nebulae'. He concluded that galaxies were moving away at a rate of 500 kilometres per second per megaparsec. This was about seven times faster than our current, more accurate, measurement, but the trend was right. The pattern would become known as Hubble's Law. Lemaître had also identified the same trend, but he would never get a 'Lemaître's Law'. Ultimately, Hubble had better data and, crucially, managed to get his message out to the wider community. With Arthur Eddington's support Lemaître did finally translate his 1927 paper into English in 1931, published in the *Monthly Notices of the Royal Astronomical Society*, but he omitted the section outlining his estimate of the expansion of space, perhaps because it was already out of date. He had been pipped to the post.

Hubble himself was not sure what to make of this observed behaviour of the galaxies. But his work quickly had an impact: in 1930 Einstein was persuaded by Arthur Eddington of the importance of Hubble's results and the following year he visited Hubble in California. The results were so compelling that Einstein completely changed his mind about the behaviour of space, declaring at a 1931 lecture that 'the redshift of distant nebulae has smashed my old construction like a hammer blow'. It had finally become clear to him that the wider universe around us was growing. He also soon declared the Cosmological Constant, brought in to make the universe static, to be his 'biggest blunder'. Einstein struck it from his equations and adjusted his mind to the idea that the universe was indeed a changing place.

This discovery has shown us that we are living in a space that is growing, which has no centre and no edges. It is growing everywhere, and everything in it is gradually moving apart, except inside the galaxies and clusters of galaxies, where gravity has won out over the relatively gentle expansion. If we now imagine winding time backwards we would see space shrinking, with all the galaxies now moving towards each other. If we wind time back far enough, every galaxy would end up right next to every other one, and farther still they would be on top of each other, all occupying the same space. Here our analogies do not work, because under normal conditions on Earth an elastic can only shrink so far. In space the gaps between objects can keep shrinking almost indefinitely.

What does it mean to have all the galaxies in the same place as each other? Well, this would coincide with the moment that we call the Big Bang, the first instant in the growth of our universe, our own Time Zero, or extremely close to zero. We will say more about that idea in the next chapter. For now, something we need to know is that in the first moments there were in fact no galaxies, not yet. Instead there were extremely densely packed fundamental particles: the protons and neutrons that are the building blocks of atoms, dark matter particles, the tiny neutrino particles and rays of light.

If we could wind time all the way back to zero, the contents of space become infinitely dense, and our laws of physics break down. So instead we typically track the behaviour of the universe from a brief moment after the growth begins, and this is the time we often refer to as the Big Bang. At this point, the universe was extraordinarily compressed but it might still have stretched infinitely far in all directions. That is one of the confusing things about infinity. Even if you compress something infinitely long so that its component parts are closer together, it is still infinitely long.

Now let us wind time forwards, starting from that point of extreme density. The growth of space begins, and everything in the universe starts moving apart. This is the Big Bang, but really it is more like the energetic beginning-of-the-expansion than a bang. It can be too easy to picture the Big Bang as a colossal explosion taking place in the middle of an empty space. We might imagine things suddenly flying out through space from some central point. This would be quite wrong. There is no central point, and nothing flies out through space. It is space itself that grows. It is true that space does seem to have started to grow with explosive suddenness. But it might be better to imagine space more like a compressed spring than some sort of bomb. When the spring is released, it suddenly grows larger.

Lemaître was the first to voice this idea that a Big Bang, a beginning to the growth of space, is an obvious consequence of our living in an expanding universe. He did so in the same 1927 paper written in French and translated into English in 1931. Lemaître called what existed before growth began the 'Primeval Atom' or the 'Cosmic Egg'. His idea gathered strength in 1929 as soon as Hubble published his findings about the galaxies' pattern of expansion. The term 'Big Bang' wasn't coined until later, in the 1940s, by an opponent of the idea. The Cambridge astronomer Fred Hoyle invented it as a term of mockery as he, Hermann Bondi and Thomas Gold developed an alternative idea to explain Hubble's observations. Known as the Steady State theory, their idea was that new matter was continually being created, so that space could keep growing without ever having had a beginning. Both these ideas would be in play for many years to come.

The proponents of the Big Bang realized that if we know how fast space is growing, we can work out when the growth began. This, then, tells us the age of the universe. On a grand scale, the calculation is like a standard maths problem from school: if someone is driving at 60 miles per hour, and they are 60 miles from home, then they must have set off one hour ago, assuming they drove at the same speed all the way. The more slowly they drive, the longer the time since they set off. The same pattern works for the growth of space. The slower that space is growing, which we can measure by the speed at which galaxies appear to be moving away from each other, the longer the time that has passed since the growth began.

As an example, if a galaxy is 10 million light-years away from our Milky Way, and it appears to be moving away from us at 30 billion miles per year, then we could work out that it would have 'set off' from our Milky Way 2 billion years ago. This is because the time taken to move somewhere is the distance moved divided by the speed of motion, so long as the speed stays the same, and the distance to the galaxy is almost 60 billion billion miles when we convert from light-years. So, a distance of 60 billion billion miles at a speed of 30 billion miles per year takes 2 billion years. In a space that

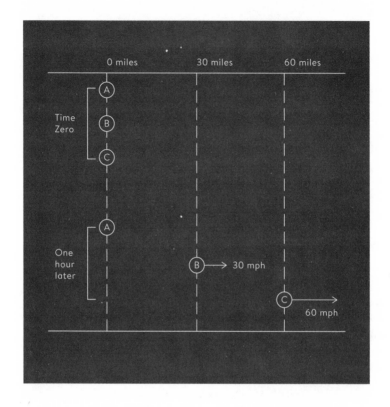

Figure 4.6

If a car driving at 30 mph is 30 miles away, and one driving at 60 mph is 60 miles away, they must have both set off an hour ago. To find the age of the universe we use a similar calculation with galaxies instead of cars.

is expanding uniformly, this time would work out to be the same for the more distant galaxies too. A galaxy twice as far away from the Milky Way would be moving away twice as fast. So too a galaxy ten times farther away would be moving ten times as fast. We would find the same answer for the time they all set off: 2 billion years, for any galaxy.

But what is the significance of this number? Well, it describes the amount of time that has passed since the expansion of space began. Since we consider our universe to have begun at the Big Bang, it is none other than an estimate for the age of the universe itself. So is our universe 2 billion years old? No, we actually think it is several times older than that, almost 14 billion years, but these illustrative numbers do in fact match the measurements made in 1929 by Edwin Hubble and Milton Humason. Their measurements would have implied that the expansion of space began 2 billion years ago, assuming that space had been steadily growing throughout its history. This was problematic at the time, as geologists were able to estimate the age of the Earth using radioactive dating of rocks. By the early 1930s the British geologist Arthur Holmes had showed that some of the rocks in the Earth were more than 3 billion years old. How could the Earth be older than the universe? Something wasn't adding up.

As we will find out in chapter 5, this estimate has to be refined to account for space not growing steadily throughout its life so far. But that was not the only problem. It later became clear that Hubble had underestimated the distances to the galaxies. This meant he had overestimated the rate of expansion of space, now known as the Hubble constant, implying a universe that was too young. In the years after 1929 astronomers continued to make more accurate measurements of distances and apparent speeds for a larger set of galaxies surrounding us. During the Second World War Walter Baade used the 100-inch Hooker telescope at Mount Wilson to study individual stars in Andromeda, aided by the reduced light pollution in the Los Angeles area during the blackouts. In 1952 he announced that there were in fact two distinct types of Cepheid stars. Accounting for these different populations halved the expansion rate and doubled the estimated age of the universe.

This work has continued up to the present day and has had a colourful history, with two prominent rivals, the astronomers Allan Sandage and Gérard de Vaucouleurs, disagreeing for many years in the latter half of the twentieth century over the size of the Hubble constant. Sandage was an astronomer at the Carnegie Observatories who had worked as Hubble's assistant while a graduate student, both at Mount Wilson and with the 200-inch Hale telescope that opened on the Palomar mountain in Southern California in 1949. He took over Hubble's programme after his death in 1953 and used the Hale telescope to measure the pulsating Cepheid stars. He estimated the stars to be at even greater distances than Baade had and by the 1970s he was convinced that the data preferred an expansion rate of only 50 kilometres per second per megaparsec, ten times slower than Hubble's original estimate. This much lower rate would have implied that the universe was as much as 20 billion years old if it were expanding at a constant rate.

Gérard de Vaucouleurs, a French astronomer based at

the University of Texas in Austin, attacked this estimate, claiming that Sandage's measurements of the distances to distant galaxies were wrong. He, instead, estimated the rate of expansion to be twice as fast, pointing to a universe that was only about 10 billion years old. Throughout the 1970s and '80s there were such strong disagreements between the two at scientific meetings and conferences that they became known as the 'Hubble wars'.

Resolution of this thorny debate came in the form of Wendy Freedman and her team. Freedman is a Canadian-American astronomer who was a staff scientist at the Carnegie Observatories, later to become their director, and in 1987 was the first woman to have been made a permanent staff member. She led the Hubble Space Telescope Key Project together with astronomers Robert Kennicutt, at the time working at the Steward Observatory in Arizona, and Jeremy Mould at the California Institute of Technology. Their team used that magnificent space telescope to better measure the distances to 800 distant Cepheid stars in galaxies reaching as far as 20 megaparsecs from Earth, and then to extend the reach out to more distant galaxies using Type Ia supernovae in galaxies as far away as 400 megaparsecs from Earth. This is over a billion light-years and reaches well beyond our own supercluster. Light arriving from these most distant galaxies would have arrived with a wavelength 10 per cent longer than when it set off on its journey to Earth.

In 2001 the Key Project team published their findings, confirming ever more precisely that the further away the galaxy from the Milky Way, the faster it recedes, corroborating the prediction of an expanding space. They concluded that the

Hubble constant is just over 70 kilometres per second per megaparsec. This new measurement was uncertain at only the 10 per cent level and landed right in the middle of the two numbers that Sandage and de Vaucouleurs had been fighting over for years. Accounting for the fact that the expansion of space has not been quite constant over its lifetime, this new result allowed astronomers to pin down the age of the universe to be almost 14 billion years, a success that would win Freedman, Kennicutt and Mould the Gruber Prize for cosmology in 2009. Other measurements of the universe would refine the precision of this time since the Big Bang still further, as we will discover in chapter 5.

For much of the twentieth century, scientists debated not *when* the Big Bang occurred, but whether it occurred at all. Opinion was deeply divided. In the mid-1930s, just a few years after Hubble announced his findings, many were already persuaded, but others, spearheaded by Fred Hoyle at Cambridge, found the Steady State theory a more compelling explanation. Progress was to come in the 1940s, driven by Russian-Ukrainian physicist George Gamow and American physicists Ralph Alpher and Robert Herman, working together at the Applied Physics Lab at Johns Hopkins University. Gamow was Alpher's doctoral adviser and would later be Vera Rubin's too. This group worked out that if the Big Bang really had happened, then there should still be light left over that was formed during those earliest moments. We should still be able to see some of it now, and Alpher and Herman calculated in 1948 that it should be a chilly few degrees above absolute zero. Absolute zero is the coldest temperature that anything can reach, a little cooler than −270 degrees Celsius.

Why should there be light around now, left over from the Big Bang? If we could wind time backwards, we would see all the galaxies in the universe get closer and closer together, until eventually arriving at a time in the deep past when there were no galaxies at all. In the earliest part of the universe's life we think there were just densely packed fundamental particles and rays of primordial light, produced in the first instants after the Big Bang. A time before stars even existed. We shall learn about how those particles ended up turning into stars and galaxies in the next chapter. Gamow, Alpher and Herman worked out that there would have been a particular epoch, when the universe was almost 400,000 years old, when these rays of light would have suddenly started streaming out in all directions from places throughout space.

Why then? We should imagine the universe starting out very hot and dense, with particles tightly packed together. As space grew and things in it had more room to spread out, the ambient temperature would have got cooler and cooler. That particular age, 400,000 years, marks an epoch when everything in space had cooled from trillions of degrees right down to a few thousand degrees. Before then, the extreme heat would have kept atoms broken into their fundamental parts of nuclei and electrons. Faced with a sea of electrons, rays of light change direction every time they reach one, so in the early years of the universe space would have been full of rays of light zig-zagging about through space in a random way. That quickly changed when the universe

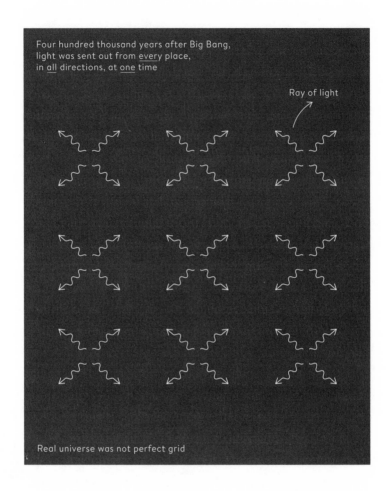

Figure 4.7
Cartoon of primordial light being sent out throughout space, like a set of lightbulbs flashing on and off.

got cold enough to let atoms become whole, their electrons captured. Neutral atoms don't cause light to swerve in their path, so suddenly every one of these light rays could set off on a straight-line journey through space.

A similar effect could be imagined by spreading lightbulbs throughout space and then briefly flashing on each lightbulb. A burst of light would come out from each of these light-bulbs, and it would flow out in all directions. These bursts of light in the real universe would have been produced by those rays of light that were spread throughout the universe and had suddenly become free to travel out in straight lines through space.

If we imagine light travelling out from a particular point in space, it will soon pass through the part of space closest to it, travelling onwards. A ray of light can travel through space indefinitely, so even though it set off only a few hundred thousand years after the Big Bang, it will now, almost 14 billion years later, astonishingly still be travelling, unless it happened to hit an object in space. Remember, most of space is empty. Right now, we humans here on Earth are being hit by the particular rays of that ancient light that set off just the right distance away from us that they are reaching us, today. This is the distance that light has been able to travel in the almost 14 billion years that have passed since the universe was a few hundred thousand years old. Which direction is this light coming from? It is coming from every direction, all around us, from places in space that are on the surface of a sphere that is centred on the Earth.

The light from those same places will have set off in other directions too, never to reach us on Earth. Light that set off from places further away will not have reached us yet, but will arrive here in years to come. Light that came from places nearer Earth will already have hit us. There is nothing special about our place on Earth: a sentient being on a planet in a different galaxy would also find themselves hit by different rays of this light, originating from a different part of space.

This light will have started its journey when the ambient temperature was a few thousand degrees, and at that time it would have had a short wavelength, about a thousandth of a millimetre, reaching into the visible part of light. As space has grown it has cooled down, and so too has the light. The wavelength of the light will have lengthened as space expanded, until today it is typically a couple of millimetres in wavelength, right in the range of microwave radiation.

This was the primordial light that was predicted back in 1948. As a consequence of the universe beginning in a hot Big Bang we should not only see galaxies receding from us, but also find ourselves bathed in cold microwave light that is hitting the Earth from all directions. In a Steady State universe, we would see the galaxies behaving in the same way, but we would not expect this bath of radiation. This was a clear way of distinguishing between these two popular scenarios, but, like early research on dark matter, this brilliant work by Alpher and Herman was to lie dormant for more than a decade. Radio and microwave telescope technology was not yet sufficiently advanced to measure this light.

In the early 1960s, Russian physicist Yakov Zel'dovich began renewed work on Gamow, Alpher and Herman's predictions for the relic light, and in 1964 his colleagues Andrei Doroshkevich and Igor Novikov wrote a paper suggesting

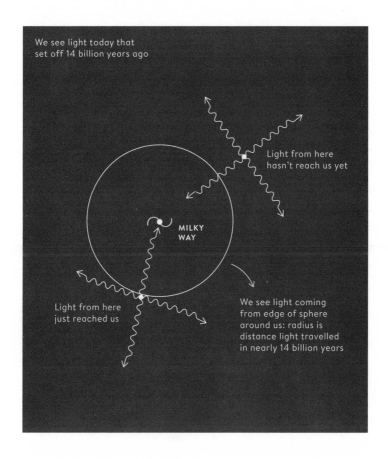

Figure 4.8
Illustration of how we receive primordial light that set off nearly
14 billion years ago, emerging from places in space that take the form
of a sphere around us.

that the radiation was a measurable phenomenon. In the United States, physicist Robert Dicke at Princeton University independently came up with the idea of an existing bath of radiation, encouraging his then post-doctoral researcher Jim Peebles in 1964 to work on the theoretical details. Peebles would come up with similar predictions to Gamow, Alpher and Herman, only later becoming aware of their earlier work. By the 1960s the technology had also caught up with theory: Dicke had invented a new radiometer, a device to measure the intensity of radio or microwave light, as part of war research at the Radiation Laboratory at the Massachusetts Institute of Technology. He designed it to pick out faint signals by switching between the real light coming from the sky and a known man-made reference signal: instrumental noise coming from inside the device would appear in both cases, so any difference between the signals should be coming from the sky. Dicke encouraged his post-doctoral researchers in Princeton, David Wilkinson and Peter Roll, to begin building a detector using a Dicke radiometer, hoping to find the relic light that would prove the Big Bang's existence.

At the same time, physicists Arno Penzias and Robert Wilson were working just down the road from Princeton, at Bell Labs in Holmdel, New Jersey. They were using a large horn antenna as a radio telescope to make some measurements of the Milky Way. Made of aluminium, this is a horn-shaped collecting device somewhat similar to a trumpet horn but much larger: this one had a square opening 6 metres across that directs radio waves in from the sky to be detected. Wherever they pointed the antenna, they kept picking up a faint hum, a tiny signal that seemed to have no source.

They did everything they could think of to eliminate the signal, including – famously – removing pigeon droppings that had collected in the antenna and could have been the source of the hum. They even resorted to shooting the pigeons, but despite all their efforts the signal was still there.

In the end, Arno Penzias spoke by chance to radio astronomer Bernie Burke, who had heard about a seminar in which Jim Peebles spoke of the expected existence of the background radiation, and the activities underway at Princeton to look for it. It seemed to be a plausible explanation for the signal they were picking up, so they called Dicke, the leader of the Princeton group. After a conversation about their measurement, Dicke shared the news with his junior colleagues Peebles, Wilkinson and Roll, famously declaring 'Well, boys, we've been scooped.' They had been, but in the end the two groups published papers together in 1965: Penzias and Wilson reported their discovery of this Big Bang radiation, while Dicke, Peebles, Wilkinson and Roll detailed the theoretical interpretation. Soon afterwards, Dicke's group detected the relic light from the roof of the Princeton geology department. The light came to be known as the cosmic microwave background radiation, at first abbreviated to the CMBR, and in later years to the CMB radiation or light. Its discovery would win Penzias and Wilson the 1978 Nobel Prize in Physics.

The discovery of this ancient light put to rest the Steady State theory in almost everyone's minds. By the end of the 1960s the prevailing view was that our universe had to have begun in a Big Bang. Some people continued to oppose it, including Fred Hoyle, but they were in a minority. The clearest way to explain the existence of this relic light was if our universe started out incredibly hot and dense, squeezed almost unimaginably tightly together.

Given the idea of a Big Bang, there are some obvious questions that we cannot help but ask, about what happened as we wind time right back to zero. Was space really infinitely compressed? Did something happen before the Big Bang? Why did space start growing at all? These are among the most fundamental questions we have about our universe, and we don't yet have answers to them. As we try to reach back to the beginning, our understanding of physics simply breaks down. We can almost get there, to within a tiny fraction of a second, but can never quite reach zero, or at least not yet.

Einstein himself had realized that the existence of matter in space tends to slow down the growth of space rather than set it expanding, so we need something new to explain why it started growing. The most popular idea right now is something called cosmic inflation, an idea that American physicist Alan Guth came up with in 1980. His idea takes us back to within the first trillionth of a trillionth of a second of our universe's growth, the very first instant of its life. If we could return to that moment, we would not find any of the familiar things that we think space is made of, including atoms or rays of light. Instead the model of cosmic inflation tells us that we would find something stranger permeating space, something Guth named the 'inflaton field'. By 'field', Guth was referring to a diffuse energy that filled space.

According to Guth's idea, this field would have started out with some energy stored up in it, a little like a compressed spring. Just as a spring bounces apart when released, so would the space filled with this inflaton field. It would have done so extremely rapidly, growing everywhere. In a fraction of a second two points spaced only an atom's width apart would have ended up more than a million light-years apart. The growth would have been exponential, which means space doubling in size every tick of the clock. The distance between points in space would have grown even faster than the speed of light.

If true, after much less than a trillionth of a trillionth of a second, the energy stored up was released, and the initial extreme growth was over. As the universe cooled down, in a process we still do not fully understand, this inflaton field would have turned into the familiar ingredients of atoms, rays of primordial light and presumably dark matter particles. Space would still be expanding vigorously, but it would now be starting to slow down as the gravity of all the material exerted a braking effect.

Guth came up with this idea of inflation to try to make sense of some cosmic oddities. In the 1970s a peculiar feature of the microwave background radiation had become apparent. The radiation was almost precisely the same temperature everywhere one looked, in every direction on the sky, even though it had been travelling for billions of years from very different parts of the universe to reach us. That would suggest that the diverse points of origin, spaced incredibly far apart, were also the exact same temperature. This should only be possible if the points had been in contact with each other at some time in the past, just as an ice cube and a glass of water have the same temperature when the one melts into the other. Guth's inflation explains that all of the parts of space that we can now see, the whole of our observable universe, were once packed tightly closed together, in contact.

We do not yet know if this idea of inflation is true. It matches up with everything we have observed (which we will discuss in the next chapter), but it remains elusive, impossible so far to nail down with certainty. If inflation did take place then it should have left its own faint but distinctive mark on space. The act of inflation should have rippled space-time, generating gravitational waves travelling through space just like the colliding black holes made the space-time ripples that were triumphantly detected by the LIGO experiment in 2015. Typically, the more energy in the inflaton field, the larger the ripples would be, and they should have imprinted a particular pattern in the microwave background light. If a gravitational wave was passing through space when the universe was 400,000 years old, when the CMB light set off, it would have distorted space, stretching it in one direction and squeezing it in another. This would have had the effect of subtly polarizing the rays of light, inducing them to vibrate preferentially in one direction more than another. On mountaintops in the deserts of northern Chile, and in the cold and desolate highlands at the South Pole, astronomers are pointing microwave telescopes to the skies, looking for these signals.

If they are discovered, it will be marvellous. We will have stepped closer to the limits of what we might ever know about our origins. Of course, we will then wonder why the inflaton field was there in the first place, with that particular amount of energy. It is also quite possible that this theory is wrong and we will not find any hints of these ripples. There are prominent physicists, including Paul Steinhardt at Princeton University and Neil Turok at the Perimeter Institute in Canada, who argue that the idea of cosmic inflation is fundamentally flawed, and that inflation would not naturally produce a universe that looks like ours if we take into account quantum mechanics properly. They argue that we need to go back to the drawing board and think harder about what else could have happened in those earliest moments of our universe's life. And they and others, including physicist Anna Ijjas at Columbia University, are coming up with alternatives to inflation, for example 'bouncing' models in which the Big Bang is not the beginning of the universe after all, but instead just a moment in a longer, perhaps cyclic, history of expansions and contractions of space.

So far we have focused on the question of whether space has always existed, but we can also consider how to answer our question about whether space is infinitely large. For a space like ours that looks on average the same everywhere and in all directions, the answer to this question depends on two defining characteristics. These are the geometry and the topology of space.

The geometry of space tells us how curvy it is. We are familiar with surfaces in two dimensions that have different geometries. The surface of a piece of paper is flat. The surface of an orange, or of the Earth, is curved. How much a surface is curved tells us what its geometry is, and you cannot change the geometry of something without deforming it. No matter how we connect it up, we can never turn a flat piece of paper into the curved surface of a ball without breaking it. The reverse is also true: we could only make the skin of a round orange into a flat surface by tearing it. Their geometries are different.

A simple way to tell if two surfaces have the same geometry is to draw a triangle on the surface. Most people learn in school that the angles in a triangle always add up to the same thing: 180 degrees or the equivalent of two right angles. But that is only true for a triangle drawn on something flat. Draw a triangle on an orange, and you will see that the angles add up to more than they would on a flat surface. An extreme triangle would start at the North Pole of your orange, and one side would come down to the Equator. The second side would go a quarter of the way round the Equator, then the third side would come back up to the North Pole again. Each of the three angles in that triangle would be a right angle, adding up to 270 degrees.

You can also draw straight lines on these different surfaces to work out how curvy they are. Two straight lines that start out parallel on a flat surface stay parallel for ever more. On a surface curved like an orange, those two same parallel lines will at some point come together. Another sort of curved surface imaginable in two dimensions is the surface of a saddle, or of a Pringle crisp. This surface is also curved, but unlike a ball it curves in two different directions, curving upwards back-to-front and downwards side-to-side. If you drew a triangle on a Pringle crisp, the angles would add up to less than a usual flat triangle. If you drew parallel lines on the crisp they would begin to diverge, getting ever further apart.

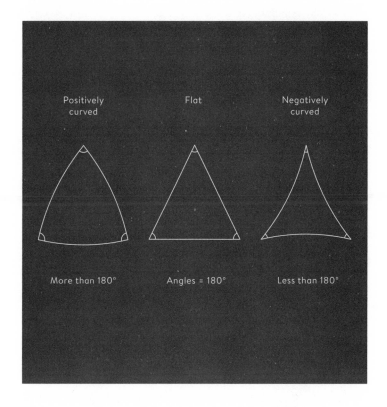

Figure 4.9
Different geometries of space.

These descriptions capture the geometry of two-dimensional surfaces, but apply equally to the curviness of our three-dimensional universe. Space can be flat, or positively curved, or negatively curved. These categories are harder to visualize, because to picture the two-dimensional surface of a ball, we think about the whole ball, sitting in three dimensions. To picture an equivalent curved three-dimensional space, we would need to be able to think in four dimensions. Frustratingly, we humans are not equipped to visualize that, but we can still imagine making measurements in three dimensions. Just like in two dimensions, if the geometry of our space is flat, then parallel lines, like light rays, stay parallel always. If the space is curved then they will either come towards each other in a positively curved universe or splay apart in a negatively curved one.

In a space that has no edges, a positively curved universe would not be infinitely big, in the same way that the surface of a ball is not infinitely big. As on the Earth's surface, you would be able to set off one way through space in a positively curved universe and, if you travelled for long enough, eventually come back around the other side. Set off in another direction and you would come back round again too. Unlike on the surface of the Earth, though, you could set off in any direction in space, in any of the three dimensions, and eventually return to your starting place.

What determines the geometry of our universe? It depends on the amount of matter in it. The more space weighs on average, the more it gets deformed, just as a lead ball would deform a rubber sheet more than a foam ball of the same size, an example we encountered in chapter 3. We can measure the geometry by measuring how deformed space is, by tracking the path of light through it. As light travels through space to us here on Earth from distant objects, it will travel on a straight path if space is not deformed. The more deformed it is, the more a light ray will curve, just like that marble on a trampoline.

I can now imagine that trampoline with either a lead or a foam ball placed on it. I stand one side of the trampoline, and on the other side I have a friend who stretches her arms out wide and rolls a marble to me from each hand at the same time. When the foam ball is on the trampoline, it barely distorts the surface, and the two marbles roll in straight lines to me. When we replace it with the lead ball, the surface of the trampoline bends, and the two marbles must take a longer, curved journey to reach my hands. If I now measure the angle that separates the path of the two marbles when they arrive at my hands, it would be smaller for the foam ball than for the lead ball. A triangle drawn in a flat geometry has smaller angles than a triangle in a positively curved geometry. This means that I can use the size of the angle between the marbles' paths to work out how much the ball weighs. In the case of the lead ball, the angle would be larger, and if I didn't know that the marbles had followed a curved path, it would appear as if my friend had extremely long arms.

In the real universe I can imagine a similar situation that will allow me to work out the geometry of space. Instead of two marbles from my friend's outstretched arms I use rays of light coming from a distant part of space, and instead of the trampoline I have space itself. If I use light that has travelled a very long way, I can use it to weigh the total mass of

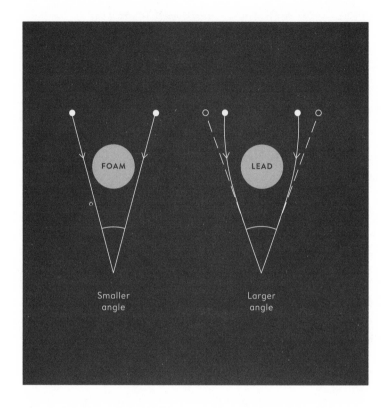

Figure 4.10
Weighing space by using the angles in a triangle.

all the matter in space that it has travelled past, not just the weight of one lump of matter in space. The cosmic microwave background light is ideal for this, as it has been travelling for almost as long as the universe has existed. Then, apart from the challenge of measuring the light, the cosmic equivalent of catching the marbles, I also have to work out what angle to measure. In the trampoline example I knew the span of my friend's arms, so I could figure out whether the triangle had angles that added up to 180 degrees.

It turns out that there are subtle features in the cosmic microwave background light, spots that are brighter or dimmer than average, that we can use in place of my friend's outstretched arms. We will find out why they are there in chapter 5, but for now it is enough to know that they appear about the size of your thumb's width held up at arm's length against the sky. The size of a typical spot makes the short side of an extremely long and narrow triangle: the length of each of the two long sides is the distance that the CMB light has travelled on its entire, near 14-billion-year journey to us. The spots will appear larger in the universe that weighs more, that distorts space more and causes the light to travel on a curved path.

There was therefore this wonderfully grandiose aim, to weigh the entire observable universe by measuring the angle of this immense triangle, and it was achieved with great success in 2000. Two competing balloon experiments flew high above the Earth's atmosphere to measure these particular features in the microwave light. Launched from the Columbia Scientific Balloon Facility in Palestine, Texas, the American-led MAXIMA experiment flew for eight hours in

August 1998. In December of that same year the American-
and Italian-led BOOMERanG experiment launched from
McMurdo Station in the Antarctic and flew for ten days. Both
teams found and measured the spots and discovered that
they were precisely the size expected in a universe in which
light travels in straight lines. They had weighed the universe,
and discovered that space, on average, does not appear to be
curved at all.

So, how much did they find that the universe weighs?
Surprisingly little. If we spread all of the matter in the universe out uniformly, then every metre-long box of space
would weigh only as much as about six atoms of hydrogen.
Of course, the universe is much denser than that in particular places, where there is a cluster of galaxies or a clump of
dark matter. But this is a good reminder that most of space
is empty.

The apparent flatness of our space tells us that it could
be infinite, stretching on for ever in all directions. Being flat,
though, does not necessarily mean that it is infinitely big. The
topology of space matters too. The topology is a description of
how space is wrapped up, and we can change the topology of
a space without deforming it at all. We can take our piece
of elastic as an example. Without deforming the elastic, we
could either lay it out in a long straight line or take the two
ends and connect them up to make a circle. We could do
something similar with a piece of paper, laying it out flat
or rolling it into a long tube. The geometry of the paper, or
the elastic, is unchanged in each case, but the topology is
different.

We can think about the universe too as having different

Figure 4.11
Some different topologies that a strip of paper can have.

topologies. Indeed, space might be rolled up in much the same way as we roll up a piece of paper, but we cannot so easily visualize it. The surface of the paper is two-dimensional, and we imagine a rolled-up tube of paper embedded in three dimensions. We would need to be able to imagine a fourth dimension if we are to connect up a three-dimensional space in the same way. It could be achieved by connecting up the left- and right-hand sides of space, the back-and-forth sides, and the up-and-down sides such that whichever direction we were to travel in through that space, we would end up coming back to our starting point.

If space were connected up like this, it would be finite in size but would not have any edges. If that size were very much larger than the observable universe, we would have no way of distinguishing it from an infinitely large space that was not connected at all. But, if the size were smaller than the size of the observable universe, we could have noticed that already. What would such a universe look like? At first glance, not necessarily so different to an infinite one.

We can take the example of the tube of paper to imagine what we might see. Here we are thinking about a space that consists just of the two-dimensional surface of the paper. Now, imagine installing an ant with a head-torch on the surface of the paper, and a second ant as the observer. Of course, ants are three-dimensional creatures, but we will think of them as flat. Next, let's assume that the light coming out from the torch doesn't stream out away from the paper, but instead is confined to travel along the two-dimensional surface of the paper. The observer will see light coming from the ant's torch. But the torchlight keeps going all the way round the tube of paper, taking one or more full loops round before again reaching the observer. That light is still coming from the ant's torch, but each time it loops around it takes a little longer to be seen and travels further.

Light will keep travelling until it hits something, so the torchlight will keep looping round and round the tube. Each time it loops around, the observer will be able to see it. The overall effect will be for the observer to see multiple images of the torchlight, with regular spacings between them that correspond to the distance round the tube. The nearest image will be the most recent, and then successively farther away images will be ever older, as the light takes time to loop round and round.

In a finite universe this effect would happen in three dimensions. The torch might now be a galaxy out in space, and we are the observers here in the Milky Way. The light from the galaxy would reach us, and we would see the galaxy. But a long time before, light from that same galaxy would have set off into space, travelled all the way round the finite universe and come back to where it started, just as if we set off on a voyage around the Earth we would return to where we started. Those rays of light would reach us too at the same time as the first ones, but they would show a much older image of the galaxy. In this way, we would get to see multiple images of the same cosmic objects as we look out into space. We would be deluded into thinking that the universe was much larger than it really was. American cosmologist Janna Levin, an expert in cosmic topology, likens the effect to being in a hall of mirrors.

Of course, we would notice it easily if the size of the

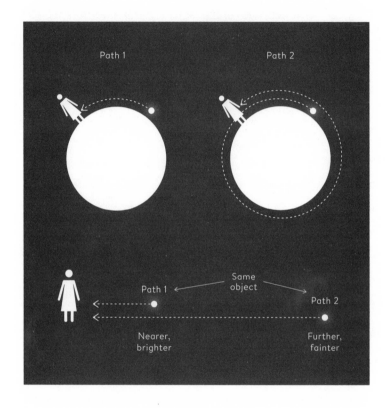

Figure 4.12
Multiple images of the same object in a finite universe.

universe were, for example, as small as our Local Group of galaxies. It would be harder to notice if the size were many times larger, as we have not yet been able to map out the farthest reaches of the observable universe even with modern telescopes. But astronomers have not yet found any evidence for these repeating patterns. This doesn't mean that the universe has to be infinitely big, but it does mean that if it is finite and connected up side-to-side, front-to-back, top-to-bottom, then the size of it must be larger than the part of the universe we can observe.

If the universe is infinite, it may not behave the same on average everywhere. If inflation did happen, it predicts that different pockets of the universe would have started growing at different times, each of them becoming its own sub-universe that could have its own laws of physics and its own fundamental particles. What we might call 'our universe' would then just be the part that inflated at one particular moment to produce the part of space we can see, the part containing us. If the Big Bang is what we call the start of the inflation of our own bubble, it is possible that the larger universe could have already been in existence before then.

This idea of many bubble universes is one possible example of something called a multiverse, the idea that our universe is just one part of a much larger conglomeration. It has strong advocates and strong opponents. Strong advocates include the Russian-American physicist Andrei Linde at Stanford University, one of the founders of inflation. The strong opponents include Paul Steinhardt, who argues that the multiverse that inflation would produce would be overwhelmingly dominated by regions which are physically different to the ones we see. Following the argument, inflation can make no predictions for how our own part of the universe should behave. This, Steinhardt argues, is one of the things that makes it a scientifically flawed idea.

If inflation is not the right idea, it is possible that the universe continues infinitely far with the same conditions throughout. This is curious to contemplate, though, because it follows that there might be infinite copies of each of us somewhere out in that vast beyond. This is a strange notion, and, to many of us, the idea of a finite universe is a more palatable option. But whether space is finite or infinite, we may never know. We have, however, learned an enormous amount about the entire part of space we can ever hope to see. We know that it is growing, and that it began to do so almost 14 billion years ago. We just don't yet know why.

CHAPTER 5

From Start to Finish

Returning now to our place in the universe, we locate ourselves on our small planet travelling around the Sun. Our Sun is surrounded in space by its neighbouring stars, many of them encircled by their own tiny planets. Our neighbouring stars move around in the longer spiralling arm of stars that makes up part of our larger home, the Milky Way galaxy. Our Galaxy, a huge disc of stars and gas embedded in a much larger halo of invisible dark matter, is spinning gently around. We look out to our neighbouring galaxy, the majestic spiralling Andromeda, slowly moving towards us through the depths of space. Around us there are many more galaxies, scattered through space and grouped together in smaller groups or larger clusters. Inside them, stars are born and die. Further out, we find more galaxies in their groups and clusters, as far as we can see. If we look far enough, we see them grouped into even larger structures, the megalopolis-like superclusters. The galaxies and clusters of galaxies are the bright lights on the backbone of the universe, the web of dark matter.

We know that the universe has not always been like this. It is not only individual stars that get born, but entire galaxies too. They have not always been there, and the stars within them have not always shone brightly. By noticing that the galaxies surrounding us seem, on average, to be moving away from us, we have worked out that our universe must be growing. Everything in space is getting further away from everything else. If we then wind time backwards, we are led to the inevitable conclusion that, sometime in the past, our whole universe must have started to grow. It had something that could be called a beginning, as we learned in chapter 4, although we may yet discover the universe to have existed long before this growth began.

In this chapter we will wind time forwards from that apparent starting point, discovering how we came to be where we are now, and finding out what might happen to our universe in the future. What makes this possible is that wonderful time-machine effect we experience from peering into space. The further you look, the further back in time you get to see. As we saw in chapter 1, by seeing other parts of space as they were in the past, we can piece together how the whole universe evolved.

Our universe, as viewed today, is not at all uniform. There are parts of space that are almost completely empty, and parts of it that might contain the richness of a solar system, the density of a black hole or a tight knot of dark matter. The initial emergence of these features in our universe had to start somewhere. If there had been no irregularities formed at the start of its growth, our universe would still be a monotonously regular sea of atoms and dark matter today. We would simply not be here.

Finding out how these initial features, which could later evolve into the cosmic objects we see today, were made is one of our biggest questions in cosmology and astronomy. It most likely happened in the very first instants of the universe's growth. The most popular explanation is the idea we introduced in the previous chapter: the almost instantaneous process of inflation during the initial trillionth of a trillionth of a second of the universe's life. During inflation, the particles that make up atomic nuclei would not yet have formed, and the whole universe would have been dominated by the inflaton field, whose stored-up energy made space briefly expand exponentially fast, doubling in size every tick of the cosmic clock.

Remember, we do not yet have any firm evidence that inflation happened, and many theorists say it is a problematic idea. But one of the reasons that the idea of inflation became so popular is that, combined with quantum mechanics, it can provide an elegant explanation for the initial creation of distinct features in an otherwise smooth space. The physics of quantum mechanics describes what happens to very small things, on the scale of atoms. As we get to those tiny scales, we start to notice strange behaviour. We find we can no longer pinpoint exactly where things are, or exactly when things happen. The energy in any given point in space can change briefly, which allows new particles to be created apparently out of nothing for a short amount of time.

If, then, we take this smooth inflaton field, its energy spread smoothly throughout space, and zoom in to examine what is happening at very small scales, we would find that tiny amounts of excess energy are constantly being created and then vanishing. If space were not expanding, this would be of no consequence. On average, on large scales, we would see nothing obvious happening at all. But in that first instant, space is growing extremely fast, so fast that space keeps doubling in size as the clock ticks. This has an important effect on those tiny amounts of excess energy. Created in a space that is growing faster than the speed of light, these quantum bumps are in almost no time at all stretched so far apart that they cannot communicate, as light does not have time to travel between them as they move away from each other. Two parts of space that started off an atom's width apart could, at the end of inflation, be light-years apart. The excess lumps of energy get stuck, embedded there in space. They cannot vanish because they have lost touch with their counterparts in now very distant parts of the universe.

This seems like a pretty weird idea. If it is true, tiny features – over-dense areas of space – are created that would normally disappear again if space were not growing extremely fast. This turns them into permanent features of space that are slightly denser than other parts. And at the end of this period of inflation, these lumps of energy turn into our familiar particles, and presumably into dark matter particles too. Where the inflaton field was denser, there would now be more particles.

These features would have been created with a vast range of different scales: some as wide across as a galaxy, some the size of a star, some the size of your hand. But they would have been incredibly subtle, denser than average by only a few parts in a million, and completely unnoticeable to the eye. Despite these modest beginnings, though, they were all we would need to set the evolution of our universe going, the tiny seeds of structure that would develop over many years

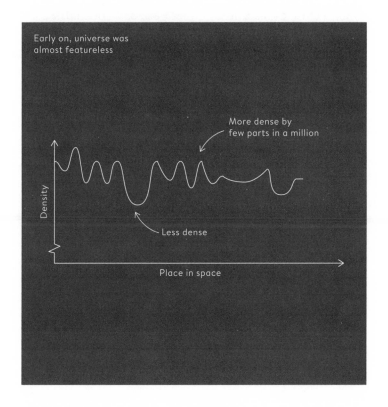

Figure 5.1
Cartoon of subtle variations in density of particles in space, imprinted during first moments after Big Bang.

to form the much more distinctive cosmic structures in our universe.

Once we leave the first fraction of a second behind, we are more confident we know what was happening. We find ourselves in a universe filled with protons, neutrons, electrons, tiny neutrino particles, rays of light and, probably, dark matter particles. It is an incredibly hot place, billions of degrees, with everything tightly packed together. It is so hot that protons can change into neutrons, and they can fuse together, forming the cores of atoms. Hydrogen fuses into helium. As time ticks past, the universe grows, and things start to spread apart. As they do so, everything gradually cools down, so that after a few seconds protons and neutrons can no longer change into each other. The number of particles that are protons as opposed to neutrons becomes fixed.

Then, after a few minutes, the whole universe gets too cold for any more fusion to happen, though it remains at a balmy billion degrees. The number of different types of atoms is now fixed, and almost all of them are hydrogen, the first element in the periodic table, made of a single proton and a single electron. For every twelve hydrogen atoms there is one helium atom, with even tinier amounts of the heavier element lithium. There is no carbon, no nitrogen, no oxygen. Not yet. To create those elements by fusing helium atoms together takes both time and heat. The universe cooled too quickly after the Big Bang to create them. We must wait until much later, when the stars with their enduring cores of extreme heat come into being.

It was the American physicist Ralph Alpher who first worked out this process for creating the primordial elements, known as nucleosynthesis. It was the topic of his PhD thesis, supervised by George Gamow at George Washington University in 1948, and the stepping stone to his prediction of the existence of the cosmic microwave background radiation. Until Alpher, observations had shown that the universe contained this particular mixture of hydrogen and helium, but no one understood why. His results were published in a paper in 1948, 'The Origin of Chemical Elements', famously signed by Alpher together with Hans Bethe and Gamow. The work was primarily Alpher's, but to his distress Gamow had included the name of his friend, the physicist Hans Bethe, to make a pun on the first three letters of the Greek alphabet. Bethe was not involved in the research, and Alpher, as a student, was rightly concerned that by writing a paper with two senior academics he would not get the credit he deserved for his breakthrough. But Alpher did get his moment of fame. Hundreds of people, including journalists, came to watch his PhD examination in 1948, with his breakthrough reported with great fanfare in the *Washington Post*.

Returning to the universe, now just a few minutes old, we would find it spread through with atomic nuclei, surrounded by a sea of tiny electrons and neutrinos, as well as rays of light and dark matter particles. The rays of light, which will later become the cosmic microwave background radiation, fly through space in all directions, changing course every time they encounter a little electron. As electrons are everywhere, the rays of light are constantly changing direction,

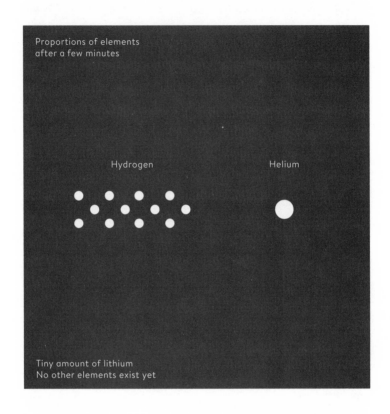

Figure 5.2
Proportions of elements formed in first few minutes after the
Big Bang.

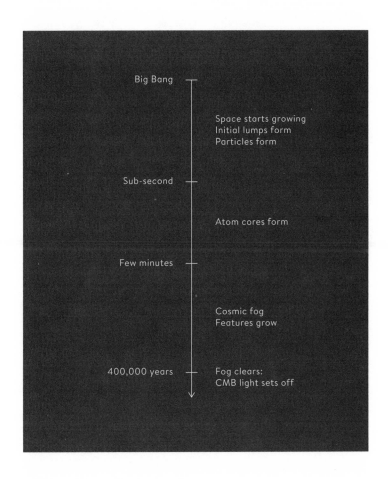

Figure 5.3
Timeline for first 400,000 years after the Big Bang.

tacking through the cosmic sea. This makes the universe behave a little like an opaque cloud or a fog.

We cannot see far through a fog, because the suspended water molecules that make up the fog bend the direction of the light rays, diverting them off their path. Any light that comes out has taken a random path through the fog. This is a little like the effect that electrons have on light in the young universe. Until it emerges from the particle fog, we cannot tell where it has been.

This fog-like universe retains the irregular features that were created in the first moments of time. These are the lumpy places where the universe is a little denser than average, where the particles are clumped together a little more. Those lumps can now begin to grow, because gravity encourages them to come together. As the pull of gravity is stronger for things that have more mass, an area of space that has slightly more matter in it will tend to pull other stuff towards it, away from emptier parts of space. Slowly, the features start to become better defined.

After about 400,000 years, the fog finally clears. During this time the universe has continued to grow, and gradually to cool down. The ambient temperature is now a few thousand degrees, which is no longer hot enough for the sea of electrons to be separated from the tiny atomic cores. Instead, entire atoms of hydrogen and helium can now exist.

Free-floating electrons can change a light ray's course, but once they are contained inside atoms, they cannot. This means that the rays of light can now travel straight through space, almost completely unaffected by the atoms and the dark matter particles around them. This light is now the

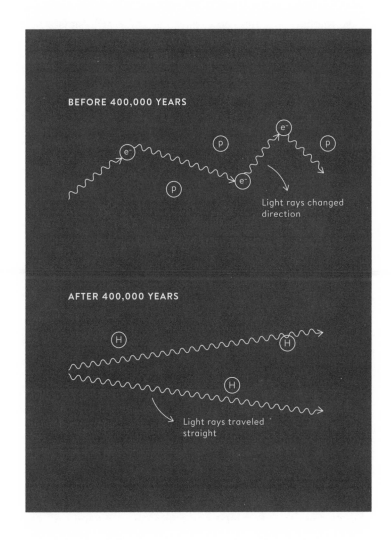

Figure 5.4
The formation of the cosmic microwave background light.

cosmic microwave background radiation that we encountered in chapter 4, first detected in 1965. The physicists Jim Peebles and Yakov Zel'dovich each led groups of researchers, in Princeton and in Moscow, who worked out in the following few years that the rays of light emanating from different parts of space should also have slightly different intensities. The intensity, or temperature, of the light would track the density of the space it had emerged from. Since different parts of space were slightly more or less dense than others, so too the light from those places should come out hotter or colder.

The effect would be subtle, though. Back at that time, the denser parts of space were only denser by about ten parts in a million, so the intensity of the CMB radiation arriving from different directions was expected to vary by only millionths of a degree. In the decades after the CMB radiation had been found, physicists turned their efforts to searching for these features. It took more than twenty years to find them, as the signal was so tiny that increasingly sensitive microwave detectors had to be developed. A good observing location was needed too, to avoid swamping the signal with contaminating microwave light from Earth's atmosphere. Eventually a team of scientists, led by the American physicist John Mather, announced their discovery in 1992 using images of the microwave sky captured by NASA's COBE satellite.

Four hundred thousand years after the Big Bang, space is now filled with the atoms of hydrogen and helium, and tiny neutrino particles, as well as dark matter particles. The regions of space that are more densely packed with atoms and particles are becoming increasingly pronounced. Gravity continues to enhance them, drawing more of the visible and invisible matter into the denser regions, and leaving the emptier spaces ever emptier. But these regions are not yet dense enough to collapse into familiar objects like stars. The universe instead enters an era known as the cosmic Dark Ages, a pre-dawn period lasting for about 200 million years when only the cosmic microwave background light lit up space. During that time the temperature of the ambient light and the atoms gradually falls from thousands of degrees Celsius to well below zero.

During this age of darkness, the denser regions begin to collapse into a cosmic network of atoms and dark matter particles, made up of structures looking like giant filaments linking up clumpy nodes. We haven't been able to see this happening so don't yet know exactly how it took place, but our understanding is advancing with our ability to create models of universes using computer simulations, like those we encountered in chapter 3. Astronomers and physicists can instruct a computer to calculate what would happen to a huge number of atoms or dark matter particles in a universe with particular features imprinted at the Big Bang. The computer needs to know about the law of gravity, and to be able to keep track of the movements and groupings of a large enough number of different objects. As our computers get increasingly sophisticated, we can simulate the universe in increasingly high fidelity.

These simulations give us confidence that we know what was happening back then, but we may never be able to peer right back into the cosmic Dark Ages. No stars existed, so there is no starlight from the period to be seen. There is a glimmer of hope, though. The hydrogen atoms themselves, warmed up by the cosmic microwave background radiation that surrounded them, emitted some radio wave light produced when their electrons switched between two different physical states. Each hydrogen atom has one single electron, and that electron has something called its 'spin', which we can think of as analogous to whether it is rotating around clockwise or anticlockwise. Quantum mechanics tells us that the electrons are not actually spinning around, but they behave a little as if they are. When the proton and electron in a hydrogen atom both spin in the same direction, they have a little more energy than when they spin in opposite directions. This means that when the electron's spin flips to the opposite direction, the energy of the hydrogen atom drops, and that energy gets sent out as light.

This light has a particular wavelength that is 21 centimetres long, similar to the wavelength of light that we use for wi-fi signals. We might be able to measure those radio waves coming from the hydrogen atoms using specially tuned telescopes, but there is a catch. The radio waves set off when the universe was much smaller than it is now, which means that the wavelength of the light would now be much longer. Its new wavelength, many metres long, is almost impossible to see here on Earth because of all the interference from man-made devices, in particular radio stations, that send out radio waves with the exact same wavelength.

To get around this problem, astronomers seeking to study these epochs have some future-looking ideas to cover a huge area on the surface of the far side of the Moon with thousands of radio antennae. One concept, known as the Dark Ages Lunar Interferometer, would be large enough to pick up extremely long wavelengths up to about 30 metres, pushing our understanding of astronomical history deep into the Dark Ages. The idea would be for an array of thousands of radio dishes to be deployed on the ground by an army of robots. Together they would be as sensitive as a single telescope a few kilometres across and would finally let us glimpse the formation of the universe in the age before stars.

After a couple of hundred million years the universe approaches the end of the Dark Ages. At last the clumps of atoms have become dense enough to form the first mini-galaxies at the dense nodes of the cosmic web of dark matter. These proto-galaxies would have been quite unlike the galaxies that we can see around us in the universe now. Many times smaller, they would have been just tens of light-years across and perhaps a million times heavier than our Sun. At first they would have contained no stars at all. By following what happens in computer simulations, we have come to think that these galaxies were each made of a disc of gas, embedded in a larger, sphere-like shape of dark matter. The ingredients of the gas would have been only hydrogen and helium, very different to star-forming gas in galaxies like our own. The ingredients of solar systems like ours, with elements like carbon and oxygen, did not yet exist.

What happened inside those mini-galaxies? The pull of gravity would have compressed the gas, heating it up to about 1,000 degrees. Where the gas was densest it would clump together ever more tightly, bringing hydrogen and helium atoms close together. Before a clump of gas can collapse into a star, though, the atoms inside it need to get cold enough for their inward-pulling gravity to win out over their outward-pushing pressure. The colder the gas, the lower the pressure. In practice this means cooling the gas clumps down to hundreds of degrees below zero, which happens when the atoms collide with each other. This slows them down, lowering their temperature, until at last the dense clouds of hydrogen and helium atoms can collapse into the very first stars. As we learned in chapter 2, fusion can then begin in their cores, generating light and heat.

Hydrogen and helium atoms do not collide and cool down as readily as gases made of elements like carbon and oxygen. This means that these earliest clumps of gas would have had a stronger outward-pushing gas pressure than we find within gas clouds in the Milky Way today. That, in turn, means that those first stars were likely born on average much heavier than a typical star today, with a stronger inward-pull from gravity to counteract the pressure. There would have been many more of the short-lived white and blue stars, the heaviest and hottest of all the stars.

We believe that the first stars formed in this way a couple of hundred million years after the Big Bang, marking the start of the 'Cosmic Dawn' of the universe. Astronomers have not yet determined the exact time this happened, because we cannot see their starlight. The stars were initially surrounded by dense clouds of hydrogen atoms which absorbed most of their ultraviolet and visible starlight, cloaking them from our view. We can, however, look for light coming from the warm hydrogen atoms themselves, that same radio wave light we mentioned earlier. Setting off with a wavelength of

21 centimetres, it should now be a couple of metres in length, shorter than the radio waves coming from even earlier in the Dark Ages. This should make the signal observable from radio-quiet locations on Earth, and astronomers are actively searching for it from isolated locations, including western Australia, the Californian desert and South Africa. They are also eagerly awaiting high-fidelity images of this radio emission from the Square Kilometre Array, due to turn on in Australia and South Africa in the 2020s. The community also have ambitions to send a new satellite, known tentatively as the Dark Ages Radio Explorer, to orbit the Moon. While orbiting around the far side, it would be shielded from man-made devices and could pick up the faint signals coming from those earliest lumps of hydrogen.

Over the next several hundred million years, the cosmic dawn broke and a transformation known as 'reionization' took place throughout the universe. The light from those first stars heated the gas around them, and the most energetic rays of light, those in the ultraviolet, would have had enough energy to break the hydrogen and helium atoms back up into nuclei and electrons, ionizing the gas. Because those stars were heavy, they would have lived short lives of only a few million years, many of them ending their brief lives as explosive supernovae. The supernovae would likely have cleared a path through the gas surrounding the stars, helping the ultraviolet light escape to break up the atoms in the gas surrounding the galaxies. Pockets of hot gas would have grown around all the galaxies, permeating the universe like holes in a Swiss cheese. They would have continued to grow over many millions of years, percolating and merging until

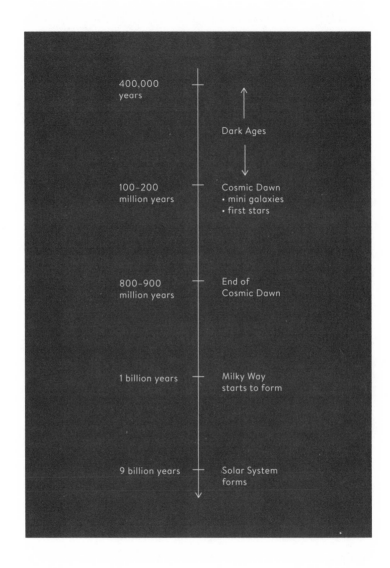

Figure 5.5
Timeline for first 9 billion years of the universe.

eventually all of the atoms throughout space were broken up into nuclei and electrons.

We still cannot say exactly how the process occurred or even when, as much of our understanding and many of our assumptions are based on computer simulations instead of actual observations. But our current estimates are that the transformation from a neutral to an ionized universe, known to astronomers as the 'Epoch of Reionization', began in earnest about 500 million years after the Big Bang, and was complete by about 900 million years. This is still approximate, but we do have some observations to guide us because we can use bright quasars as beacons. Far brighter than entire galaxies of stars, quasars are the most distant and thus the oldest objects that we can see. The American astronomers Jim Gunn and Bruce Peterson worked out back in 1965 that if you split the light from a quasar into a spectrum of all its wavelengths, tiny bits of neutral hydrogen lying in the path of the quasar's light will cut out the light at specific wavelengths. As neutral hydrogen ceased to exist after the period of reionization, it follows that if you see the quasar's light being extinguished in this way, you know you must be looking back to a time when the reionization transformation was not yet complete.

If you can then find slightly nearer quasars whose light is not extinguished by neutral hydrogen, you know you are looking at objects whose light set off just after the universe became reionized, when there was no neutral hydrogen left. The astronomer Robert Becker from the University of California led a team using data from the Sloan Digital Sky Survey to look for bright quasars just like this. In 2001 they found the first quasar whose light still showed the telltale signs of interaction with neutral hydrogen. They worked out the age of the quasars' light by looking at its redshift, determining how much its light's wavelength has lengthened between setting off and being measured here on Earth, and found it to have set off a little less than 900 million years after the Big Bang. Astronomers have also determined that light that set off from quasars a little later than this epoch shows no traces of interaction with neutral hydrogen. This dividing line seems to mark the very end of reionization, an important marker on our cosmic timeline.

There are still many open questions about that earlier part of our universe's life. We are still eager to know when the cosmic dawn started, and how it unfolded over those hundreds of millions of years. As well as measuring the starlight indirectly by viewing the neutral hydrogen using these radio wave surveys, in the next decade we hope to find many more of these distant quasars with the James Webb Space Telescope, the successor to the Hubble Space Telescope, which is due for launch in 2021. The James Webb will be NASA's new flagship mission, a multi-billion-dollar satellite with many astronomical goals that has been years in the design and construction. Its mirror is more than 6 metres across, which will give it the ability to see in exquisitely high definition, but, regrettably, make it impossible to fit inside the rocket launcher fully laid out. The mirror has thus been made in segments which will elaborately unfurl after launching. Those will be nail-biting moments for the engineers and scientists.

The James Webb's mirror will have seven times the area of Hubble, and will be perched above a huge, tennis-court-sized sunshield to keep the telescope cold. It will fly a million miles from Earth to a stable point where it can orbit the Sun and it will view infrared light as well as visible light of the longer red wavelengths. Ultraviolet starlight emitted from those first quasars and galaxies would have lengthened in wavelength many times as space grew during the more than 13-billion-year journey, arriving here on Earth as infrared light. James Webb will at last be able to see it.

Moving on in time, by the time our universe was a billion years old the gas in and around the galaxies was heated up, and the tiny galaxies had also begun to merge together to form slighter larger ones. This again was the effect of gravity as it pulled dense regions towards each other until they collided and became new and larger objects. These would have been the first objects recognizable as the kind of galaxies we know today, including one or more that would go on to form our own Milky Way. We know that part of our Milky Way must date from this time because we see stars in it that are at least 13 billion years old. Since the whole universe is just under 14 billion years old, those ancient Milky Way stars must have been part of one or more of those earliest galaxies.

Revealing how our Milky Way formed is of huge importance and interest to astronomers, but so far we only have part of the picture. We can never go back and see our own Galaxy as it was in the past; the light we observe from stars within the Milky Way set off at most about 100,000 years ago. The best alternative is to take a stab at what must have happened by observing other, more distant galaxies, whose light set off on its journey towards us millions to billions of years ago. By doing that, and by generating computer simulations that aim to model what was happening, we have become fairly certain that the Milky Way was formed from smaller galaxies merging together in the past.

As moving objects come towards each other and merge, they will tend to form a larger object that spins. When the Milky Way formed, the visible part of the merging minigalaxies, made of stars and gas and dust, came together as a spinning disc, while the invisible dark matter stayed more spherical. After another 3 billion years had passed, we think the Milky Way was more or less its current size, its spiral arms in place. The universe would have been about 4 billion years old at that point. Our Galaxy is still gently growing even now, its gravitational pull attracting smaller neighbouring dwarf galaxies to merge with it. Our two nearest neighbours, the Small and Large Magellanic Clouds, are likely to end up merging with the Milky Way in about 3 billion years, consumed by their larger companion.

Most of the stars in the Milky Way formed when it was just a few billion years old, back when it first reached its current size. The conditions then were ideal for the birth of new stars, as the merging of galaxies created massive collisions of gas clouds. Right now, there is relatively little activity in the Milky Way, and new stars are being born only very slowly, with only a few new ones appearing every year among the 100 billion stars that already exist. At the peak of its activity, our Galaxy would have made a few hundred every year.

Our own Sun was born when the universe was about 9 billion years old, towards the end of the period of peak star-creation activity. Condensing from a gas cloud deep within one of the spiral arms of stars, it would have begun life as a star with a swirling disc of dust around it. As we learned in chapter 1, planetary scientists think that our planets condensed from that dust cloud, as rocks clumped together to make ever bigger boulders, until tiny pre-planets formed that had their own pull of gravity, and eventually they became larger. We cannot ever see how our Solar System formed, just as we cannot see the deep history of our Galaxy. Instead, we can look at how other solar systems have formed around other stars and use computers to simulate what we think happened to form ours.

The Milky Way probably has a standard history for a spiral-shaped galaxy, which means it has had a less disruptive past than elliptical or oval-shaped galaxies. Those galaxies, we believe, started out as two or more spiral galaxies of around the same size. At some point, the galaxies will have dramatically collided and undergone what astronomers call a 'major merger'. A collision of galaxies on that scale would be a curiously uneventful experience to a small observer within one of the galaxies. If the Milky Way were to collide with another galaxy right now, we probably would not feel its effect. Our Solar System is too small compared to the whole galaxy, and the gaps in between stars are too vast. If you pushed two galaxies together, you would not expect any of the stars actually to hit each other, although an observer on Earth might notice a beautiful change in our night sky. Remember, in our model, that if we scaled down our Solar Neighbourhood to fit in a basketball court, our entire Solar System would be the size of a grain of salt, almost imperceptible within that much larger space.

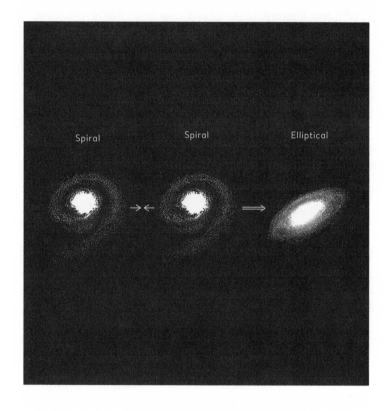

Figure 5.6
Spiral galaxies collide together to create a larger elliptical galaxy.

Even if the collision of galaxies does not immediately affect individual stars – or dark matter, assuming it is made up of small particles – effects do occur. The gas and dust that occupies the gaps between stars is spread out thinly and relatively more evenly than the stars are, and the gas clouds from the colliding galaxies combine to form new stellar nurseries, making large quantities of new stars and planets. Using up the supplies of available gas, the new supergalaxies then create very few new stars after this first burst of productivity.

The gravitational pull of each incoming galaxy also disrupts the delicate balance of its partner, sending stars into random orbits as they leave their original orbits around the original galaxy's spiral disc. The spiral shape of each galaxy disappears, and the resulting shape of the combined galaxies is likely to be spheroid: the shape of an American football, or a rugby ball, or an M&M sweet. Exactly what shape will depend on the original sizes of the incoming galaxies and the direction they were moving when they collided. Astronomers predict the outcomes of these mergers by smashing spiral galaxies together using computer simulations, finding that they routinely produce spheroid-shaped galaxies.

We are lucky enough to have captured many stunning images of spiral galaxies in the midst of mergers just like this, many of them taken using the Hubble Space Telescope or with large optical telescopes in Chile and Hawaii. Often we can see the perfect spirals just beginning to be distorted by the interaction with their neighbours. If an astronomer were to be able to take another look in many millions of years, the spirals would be gone and a single oval-shaped galaxy would remain. A merger like this is to be our own eventual fate, as the Milky Way is on its collision course with our neighbour Andromeda. Our future partner is larger than we are, but not by much, so after we do collide in just over 4 billion years, we expect to turn into a new, elliptical and super-sized galaxy.

A signal that we are eagerly awaiting should come from the giant black holes at the cores of the galaxies. These black holes, weighing millions to billions of times more than the Sun, are thought to merge together when their galaxies merge, creating ever-larger ones at the core of the new supergalaxy. When they do merge, they should produce gravitational waves, the ripples in space-time that we met in chapter 2. A gravitational wave coming from a distant merger of two giant black holes will then stretch and shrink space-time as it passes through the Milky Way.

How can we detect this? The typical separation between two of these giant black holes as they orbit each other before merging together is much larger than for black holes created from single stars, so we won't be able to detect these gravitational waves with the LIGO experiment. Instead, astronomers are monitoring pulsars – the rapidly spinning neutron stars discovered by Jocelyn Bell Burnell – in our Galaxy. They are using pulsars that spin hundreds of times per second and have timed their regular pulses precisely using radio telescopes. As the gravitational wave travels past, time itself will briefly pass differently. The effect will be to lengthen or shorten the duration of the pulses, so the goal is to look for changes in the pulse-time for a set of pulsars spread through the Milky Way. This work is underway by a team of astronomers using the International Pulsar Timing Array, which brings together groups from North America, Europe and Australia.

They are using radio telescopes across the globe, including the huge 300-metre Arecibo Observatory in Puerto Rico, the 100-metre Green Bank Telescope in West Virginia in the US, the 80-metre Lovell Telescope at Jodrell Bank in the UK, and the 64-metre Parkes Telescope in Australia. Carefully tracking a set of about sixty well-studied pulsars, within the next decade they expect to have enough sensitivity to discover the first signal from these huge black holes.

Leaving these merging galaxies behind, if we were to now zoom out to take in a wider view, we would see the whole cosmic web forming and evolving over the almost 14-billion-year history of the universe. We would see, from the time of the earliest galaxies, how gravity kept drawing nearby galaxies towards each other to form the first galaxy groups and clusters, while the network of dark matter created denser nodes and filaments, and ever emptier gaps in between them. But not all galaxies will merge. As we learned in chapter 4, even though nearby galaxies are drawn towards each other, the gentle expansion of space tends to spread out galaxies on average, so we would see their separations increasing as time passes.

Until quite recently, astronomers had long expected the pull of gravity eventually to slow the expansion of space. This assumption was based on the idea that whatever process started the expansion of the universe had long ago come to a halt. Then, the gravitational pull, whether from visible or invisible matter, should necessarily act to draw points in space towards each other rather than pushing them apart. The effect would be to slow the expansion of space.

The analogy often used here is to imagine throwing a ball in the air. You give it an initial upwards speed, and after that you let gravity do its job. What happens next depends on the speed you threw the ball. In all realistic cases the ball will slow down to a halt and turn and fall back to Earth. You can also imagine throwing it hard enough such that it would keep slowing down but would never turn back around, coasting out into space. The thing that causes the ball to slow down, and to turn around, is the gravity of the Earth pulling the ball back. If the Earth weighed much less, the ball could more likely escape its pull, never to fall back to Earth. When thinking about space, throwing the ball is a little like the initial expansion of space at the Big Bang. The speed of the ball is then like the speed of expansion of space. If it turns around and comes back down, that would be like space starting to shrink. The gravity of the Earth pulling the ball is like the gravity of all the matter in space that tends to slow down its expansion.

In the late twentieth century, one of the big questions in the minds of astronomers and physicists was therefore whether the matter was dense enough in the universe, with enough gravitational pull, to completely halt its expansion. If that were to happen, the expansion of space would not just stop but actually reverse. Galaxies would begin to move towards each other, and sometime in the distant future there would inevitably be a Big Crunch, when all of space was squeezed back down to a highly condensed, or infinitely condensed, point. In many ways this idea was appealing, as it would suggest the universe has a cyclic nature. A universe that had a Big Bang and then a Big Crunch might undergo the same process over and over again. Expansion and contraction, out and then in again. Cyclic processes are found everywhere in nature, and humans are perhaps predisposed to find them appealing.

To figure out whether there was enough mass in the universe to halt expansion, the key first task was to work out how much the universe weighs throughout space on average. As we learned in chapter 4, the more matter there is in the form of stars, gas, dust and dark matter, the denser it is, and the more gravity acts to slows down space's growth. A universe that is emptier and lighter on average will do a poorer job at slowing space down. It might slow down the growth just a little. There is a 'critical density', which is exactly the right amount of matter to slow it enough eventually to halt the growth of space, but not quite enough to reverse the expansion. A universe with this critical density would also be geometrically flat, like the three-dimensional version of the surface of a piece of paper that we met in Chapter 4. In such a universe, light travels in parallel lines.

Astronomers and physicists worked out that the critical density of the universe matched up to the equivalent of having about six atoms of hydrogen per metre-long box of space. A universe denser than this would eventually collapse to a Big Crunch. A lighter universe would keep growing for ever, albeit slower and slower. These options for the universe's future were taught as standard in schools and at universities right up until the early 2000s. All that was required was that a measurement be made of how much everything in space weighed on average.

The idea that perhaps our universe had this perfectly balanced critical density was an attractive one by the 1990s, as it was what the popular inflationary theory was generally thought to predict. It was theoretically appealing in many ways. But there were certainly some hints of things not quite adding up. By studying the positions of 2 million galaxies photographed across a tenth of the sky using the Automated Plate Measuring survey, British astronomers George Efstathiou, Steven Maddox and colleagues concluded in 1990 that on average the density must be less than half the critical density. There was also a curious observation that the age of the universe was not making sense. The expansion rate of the universe seemed to indicate to astronomers that the Big Bang happened less than 10 billion years ago. But by looking at the ages of stars, it was clear that some of them had been born more than 12 billion years ago.

In the late 1990s the two cosmic microwave background balloon experiments that we met in chapter 4 were being readied to weigh the universe more precisely. Before they had collected the data, though, two teams of astronomers reported fascinating results from measuring distant bright Type Ia supernovae. These were the kinds of supernova we met in chapter 1, the particular type that astronomers use to measure the largest distances in space, likely created when a white dwarf star gained mass from a companion star. Using a set of sensitive telescopes, two competing teams collected a new set of these supernova observations through the 1990s. One of the groups, the High-z Supernova Search Team, was led by Brian Schmidt at Mount Stromlo Observatory in Australia. The team of twenty astronomers used supernovae discovered with the 4-metre-sized Blanco Telescope in Chile, finding supernovae whose light set off as long ago as 7 billion years, when the universe was half its current age. The second team was the Supernova Cosmology Project, a group of about thirty people led by Saul Perlmutter at the University of California in Berkeley who used telescopes both in Chile and in the Canary Islands to find supernovae. Once found, the teams checked them using the Hubble Space Telescope and the largest telescopes on Earth, the 10-metre-sized Keck telescopes in Hawaii.

Rather than use these supernovae to weigh the universe, the teams were aiming to see how much the expansion rate of the universe was slowing down. In chapter 4 we encountered the idea of Hubble's Law, which states that all galaxies are on average moving apart from each other, with the more distant ones appearing to recede faster from any given observer. The expansion rate of the universe is simply how fast the galaxies are receding, given their distance from us. We can work out the distance using the supernova's brightness, and how it changes over time after the initial explosion, and we can find the apparent speed using the redshift of the supernova's light. To see how much the universe is slowing down, we would compare the expansion rate measured with very distant galaxies to the rate measured with nearer galaxies. The light from the very-distant ones set off long ago, and so they reveal to us the expansion rate in the past. The light from the closer ones set off more recently, so they tell us the expansion rate 'now'.

Both supernova teams were expecting to measure a faster expansion rate in the past, because almost everyone assumed that the expansion of space should be slowing down. The main question was simply how much it was slowing down.

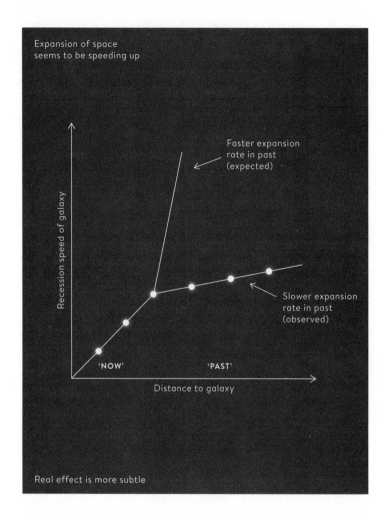

Expansion of space
seems to be speeding up

Faster expansion
rate in past
(expected)

Recession speed of galaxy

Slower expansion
rate in past
(observed)

'NOW'   'PAST'

Distance to galaxy

Real effect is more subtle

Figure 5.7
Cartoon showing how we found out that the expansion of space is
speeding up: the recession speed of distant galaxies tells us about the
rate in the past, nearby ones give us the current rate.

What they found was extremely surprising. When both teams studied their data carefully, they came to the same conclusion. They discovered that the expansion rate of the universe was apparently slower in the past, not faster. The growth of the universe was speeding up. This was as strange as if you threw the ball into space, and instead of coming back down or coasting to a halt, it instead sped up away from your hand, faster and faster away from the Earth.

The teams announced their results in 1998 to the great excitement of the astronomical community. This was a fascinating find, because nothing that was already accounted for in the universe should be able to speed up the growth of space. Well, almost nothing. Everything in our cosmic inventory that includes normal atoms, dark matter, neutrinos or light, has a pull of gravity that tends to slow down the expansion of space. One of the only things that could make space grow would be Einstein's strange Cosmological Constant, also known as Lambda. This originally was the fudge factor that Einstein put in his equations of general relativity to balance the inward-pull of gravity so that he would not have to conclude that space was expanding at all. He quickly removed it after he saw Edwin Hubble's observations of distant galaxies.

In the few years before the supernova discovery, astronomers, including George Efstathiou at Cambridge, had experimented with putting Einstein's constant back in. This made all the observations make better sense, but not everyone had bought into the idea. Now with these new supernovae results, the idea became more compelling to the wider astronomical community. Einstein's constant seemed to be back. This constant describes the energy of empty space itself. If you take a box of space that is seemingly empty, it is possible that it has some amount of energy that we call a 'vacuum energy'. This energy would have the property of making space grow faster. It would behave somewhat similarly to the energy that could have driven the initial expansion of space.

The measurements by the supernova teams showed that about two-thirds of the energy in the universe today seems to be made up of this vacuum energy. Added to the normal matter and dark matter, this also ended up making just the right amount of stuff, or energy, to give the universe the perfect critical density needed to make it geometrically flat. This was the missing piece of the cosmic puzzle: a picture was starting to come together.

In October 1998, after the results were reported, Jim Peebles and American cosmologist Mike Turner took part in a follow-up of the Great Debate in 1920 between Harlow Shapley and Heber Curtis. Debating in the very same auditorium, in the Smithsonian Museum of Natural History in Washington, DC, their new topic was 'Great Debate: Cosmology Solved?' Mike Turner argued that 'yes', cosmology was solved, Einstein's constant was back, and the universe had critical density. Jim Peebles argued 'no', that more work and observation were needed to be convinced that Einstein's Cosmological Constant was really necessary.

The results from the microwave balloon experiments would further support this picture of a geometrically flat universe, and even more evidence came in 2003 from NASA's Wilkinson Microwave Anisotropy Probe (WMAP), the successor

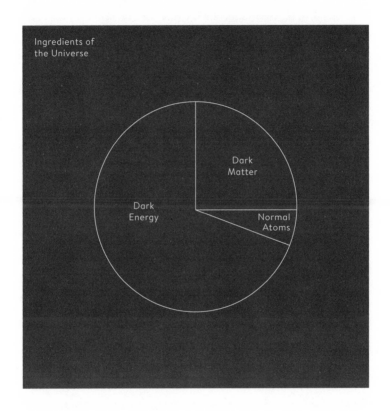

Figure 5.8
Current ingredients of the universe.

to the COBE satellite. Designed to measure the features in the CMB radiation with much greater fidelity, it was named for the Princeton physicist David Wilkinson, who had measured the radiation with Bob Dicke in 1965, and was central to the mission but who died in 2002 while the satellite was still observing the sky. The team of scientists working on WMAP, led by Charles Bennett at the Goddard Space Flight Center, concluded that the features seen in the subtle temperature variations of the CMB radiation could only be explained if we have a universe with both dark matter and a vacuum energy. They too found that the vacuum energy makes up about two-thirds of the energy in the universe today.

Within just a few years, a new reality was apparent. Our universe's growth is speeding up, and in the last third of its life so far its energy appears to have been dominated by this vacuum energy, or Cosmological Constant. Astronomers also refer to this stuff as 'dark energy', a term coined by Mike Turner, because we are still not completely sure whether it is a pure vacuum energy or perhaps something else, some new ingredient of the universe. Dark energy is a broader name for anything that could explain this accelerating growth of space.

One of the theoretical problems with dark energy is the amount of it: it doesn't yet make sense. Some possible explanations for why empty space might have its own energy relate to the rapid creation and elimination of particles following the laws of quantum mechanics. But these explanations lead to the conclusion that either our universe should be entirely dominated by this vacuum energy or its effect should be zero. There is currently no good theory for why it should have roughly the same importance, in terms of how much of the energy budget in the universe it makes up, as the visible and invisible matter in the universe. It is a huge outstanding question.

The existence of this stuff does not, however, need to worry us too much. Dark energy has no obvious effects on us in our Solar System or even in our Galaxy. It has more indirect effects, like inhibiting the clumping-together of big clusters of galaxies, because its tendency to stretch space out ever faster prevents gravity from gathering up larger and larger cosmic objects. While we do not know what it is, it is like a worrying itch, an indication that something bigger might be wrong with our whole concept of space and the laws of physics that describe it. A number of physicists are working on the possibility that Einstein's law of gravity is not quite complete, that it might, for example, need refinement to be able to explain how matter behaves when gravity's pull is very weak. It is possible that a refinement like this could be mimicking the effects of dark energy.

To help solve these puzzles, an extensive programme is underway, ready for the 2020s, to better measure how fast the universe has been growing throughout its history, and how fast the big cosmic structures have been growing. One of the telescopes designed for this quest is the Large Synoptic Survey Telescope, also known as LSST, being constructed in Chile and due to start looking at the sky by 2021. LSST will be truly vast. The whole telescope reaches as high as a five-storey building and has an 8-metre mirror to collect the light. Its camera will measure visible light and will be as large as a car, its 3 billion pixels the most ever built into a single digital camera. This giant of a telescope will scan the entire sky visible from Chile every few nights and will be able to see billions of galaxies. It will be incredibly valuable not only for understanding what dark energy could be, but also for spotting cosmic events in the sky that change day to day, like vigorous star explosions. It will help us get to know our universe ever better.

All of our observations of the sky have told us that our moment in time, here for us humans on Earth, is almost 14 billion years into our universe's history, and about 5 billion years into the Solar System's history. We now have a well-formed idea of when and how our planet got to be here, but it is also in our nature to wonder what will happen in the future. Luckily, the laws of physics are reliable enough to make pretty good predictions. We can be reasonably sure of some things. In the next hundred years or so there should be a supernova that explodes in our own Galaxy. It will be magnificent to watch and will be extremely valuable for understanding the inner workings of these exploding stars. At some time in the next few hundred thousand years the red star Betelgeuse in Orion will also become a supernova, lighting up the sky for days or weeks. Some of our lucky descendants might be around to see it.

In about 200 million years the Solar System will complete another orbit around the Milky Way. By then the familiar constellations in Earth's night sky will have changed. Many of the same celestial neighbours will still surround us, but the Sun will not stay in an absolutely fixed position relative to the backdrop of these nearby stars. It is certainly possible that during that time the Earth will also get hit by large rocky objects travelling through the Solar System.

Further ahead, in 4 or 5 billion years, our Sun will eventually run out of fuel at its core and will swell up to become a red giant. It will later end its days as a white dwarf. On a similar timescale the Milky Way will also first consume the Magellanic Clouds and then collide with Andromeda to create a brand new elliptical galaxy. Earth will be little affected by the collision with Andromeda, but strongly affected by the growth of the Sun. It will either be absorbed into the larger Sun or will be perilously close to the edge of it. It will be a seriously inhospitable place. Even before then, the Sun will become hotter over its lifetime, limiting the options for life on Earth.

We do not yet know if, in the much more distant future, the universe will keep growing. At the moment it appears as if it will do, with all the galaxies separating on average ever further apart from each other. A time may come in the very distant future when an astronomer sitting in a galaxy like ours will no longer be able to see any other galaxies, as they will all have vanished from view, disappearing over the cosmic horizon as space grows and grows ever faster. Happily, that time has not yet come, and the universe is still very much within our reach.

Looking Forward

Our community of astronomers has come an enormous way in advancing our understanding of our universe and our place within it. It is extraordinary to think that a century ago we did not even know that there were other galaxies beyond our own, we didn't know how stars created their light and we were not aware that space is growing. Even in the past twenty years we have transformed our understanding of such basic matters as the age of the universe, the nature of solar systems around other stars and the fundamental ingredients of the universe. We can now trace the evolution of the universe from the earliest moments through its almost 14-billion-year history, understanding how galaxies, stars and planets like ours came to be. Our understanding of how things work in space has taken leaps forward, allowing astronomy to evolve from a science based mainly in empirical observation into a science grounded in our deeper understanding of the physical behaviour of the objects and phenomena we see in the sky.

This is a golden age for astronomy, full of interest and possibility. One of the great excitements is that there are undoubtedly new discoveries just around the corner. Discoveries of new planets will continue apace, and perhaps soon there will be signs of conditions that hint at the possibility of extraterrestrial life. In the next few years we will no doubt see many more gravitational wave signals coming from black holes and neutron stars colliding throughout space, giving us a new way to see and understand the universe. We hope to soon discover what the invisible dark matter particles really are. And in the coming years we expect to at last see the first galaxies that formed in the universe.

These discoveries are being made possible with magnificent new telescopes coupled with ever-increasing computing capabilities. The telescopes being prepared for the next decade span all of the wavelengths of light, as well as gravitational waves, and they will target high-definition views of particular objects as well as broad surveys of the entire sky. Highlights include the Square Kilometre Array to measure radio waves, the James Webb Space Telescope to examine the infrared and the Large Synoptic Survey Telescope to map the skies in visible wavelengths. To interpret the data, our computers will continue to increase in speed and capacity, allowing ever better simulations of the cosmos and the objects within it.

There will also be discoveries that are not just around the corner, which will take much longer to reach. Being able to observe a planet suitable for life in great detail could take decades. So, too, will compiling a complete history of how our Milky Way was created. Understanding why the universe is growing ever faster, and how it started growing in the first place, will likely be a long process. But we can contemplate working towards each of these goals, because doing this work is a continual process that each of us plays a small part in. We stand on the shoulders of our scientific predecessors, all of whom have contributed in some way to the scaffolding that holds us up and that lets us together climb up further.

When we look to the future, we hand our tools and knowledge on to our students, and we plan for things that might happen fifty or a hundred years from now, anticipating the success of those who follow in our footsteps. Our past is strewn with examples of visionary astronomers and physicists who did not make the discovery that they dreamed of. Halley never got to see the transit of Venus. Hale never got to see his magnificent telescope completed. Zwicky never saw a gravitational lens. But these were not failures. These scientists inspired younger generations to keep following their path and equipped them to make their own new discoveries.

While we strive towards new discoveries, our past experience also tells us that our bigger picture of the universe and the laws of nature may still need some major adjustments. Our observations are certainly real, and our current interpretation of them tells a consistent story, but we should reasonably assume that some future shifts in the big picture are yet to come. The most exciting discoveries are the ones we least expect, ones that can radically change what we thought was true and ultimately lead us to a better understanding of our wider world. We look forward to them with eager anticipation.

Educational Resources
and Further Reading

EDUCATIONAL RESOURCES

A number of ideas in this book were developed with the help of
NASA educator Lindsay Bartolone and teacher Ilene Levine for a
NASA-funded course, 'Our Place in Space', run in 2008 and 2009.
This was part of Princeton University's QUEST professional
development programme for teachers, run by the Program in
Teacher Preparation. Those courses drew on materials developed
by educators from the Universe Forum at the Harvard-Smithsonian
Center for Astrophysics, and by many other science educators.
Particular examples are detailed below.

— The idea of scaling down the Solar System, described in Chapter
   1, originated in Guy Ottewell's *The Thousand-Yard Model*
   (*Universal Workshop*, www.noao.edu/education/peppercorn).
— The concept of scaling down increasingly large 'realms' of the
   universe into large room-sized spaces, described in Chapter 1,
   is drawn from the *Realms of the Universe* activity developed by
   the Universe Forum (www.cfa.harvard.edu/seuforum/mtu).
— The idea of simplifying the stars into four types, described
   in Chapter 2, comes from the *Life Cycle of Stars* activity
   developed by educators at Adler Planetarium for their 2001
   Astronomy Connections: Gravity and Black Holes curriculum.
— The idea of modelling the universe as a long piece of
   elastic, described in Chapter 4, comes from the *Modeling the*

Expanding Universe activity developed by educators at the Harvard Smithsonian Center for Astrophysics for the Cosmic Questions Educator's Guide (www.cfa.harvard.edu/seuforum/mtu/).

REFERENCES AND FURTHER READING

Books

Barrow, John, *The Book of Nothing* (London: Vintage, 2001)

Begelman, Mitchell & Martin Rees, *Gravity's Fatal Attraction: Black Holes in the Universe* (Cambridge: Cambridge University Press, 2009)

Close, Frank, *Neutrino* (Oxford: Oxford University Press, 2012)

Coles, Peter, *Cosmology: A Very Short Introduction* (Oxford: Oxford University Press, 2001)

Ferguson, Kitty, *Measuring the Universe* (London: Headline, 1999)

Ferreira, Pedro, *The State of the Universe* (London: Weidenfeld & Nicolson, 2006)

——, *The Perfect Theory* (London: Little Brown, 2014)

Freese, Katherine, *The Cosmic Cocktail* (Princeton: Princeton University Press, 2014)

Haramundanis, Katherine (ed), *Cecilia Payne-Gaposchkin* (Cambridge: Cambridge University Press, 1984)

Harvey Smith, Lisa, *When Galaxies Collide* (Melbourne: Melbourne University Publishing, 2018)

Hawking, Stephen, *A Brief History of Time* (New York: Bantam Books, 1988)

Hirshfield, Alan, *Parallax* (New York: Freeman & Co., 2001)

Johnson, George, *Miss Leavitt's Stars* (New York: W. W. Norton & Co., 2006)

Lemonick, Michael, *Echo of the Big Bang* (Princeton: Princeton University Press, 2003)

Levin, Janna, *Black Hole Blues* (New York: Alfred A. Knopf, 2016)

——, *How the Universe got its Spots* (London: Weidenfeld & Nicolson, 2002)

Miller, Arthur I., *Empire of the Stars* (London: Abacus, 2007)

Panek, Richard, *The 4 Percent Universe* (Boston: Houghton Mifflin Harcourt, 2011)

Peebles, P. James, Lyman Page & Bruce Partridge (Eds), *Finding the Big Bang* (Cambridge: Cambridge University Press, 2009)

Sobel, Dava, *The Glass Universe* (London: Penguin, 2016)

Tyson, Neil deGrasse, Michael Strauss & J. Richard Gott, *Welcome to the Universe* (Princeton: Princeton University Press, 2016)

Weinberg, Steven, *The First Three Minutes* (New York: Basic Books, 1993)

Wulf, Andrea, *Chasing Venus,* Knopf, 2012

Selected journal articles

Many of these articles are freely accessible online, and can be found using the SAO/NASA Astrophysics Data System Digital Library (adsabs.harvard.edu/abstract_service.html), or via the arXiv e-print service (arxiv.org). They are listed in order of appearance in the book.

CHAPTER 1

'A new method of determining the Parallax of the Sun', E. Halley, *Phil. Trans. R. Soc. Lond.,* Vol XXIX, No 348, 454 (1716) (p36 reference, translated from Latin)

'A low mass for Mars from Jupiter's early gas-driven migration', K. Walsh et al., *Nature,* 475, 206 (2011)

'Discovery of a Planetary-sized Object in the Scattered Kuiper Belt', M. Brown, C. Trujillo & D. Rabinowitz, *Astroph. Jour.,* 635, 97 (2005)

'Evidence for a Distant Giant Planet in the Solar System', M. Brown & K. Batygin, *Astroph. Jour.,* 151, 22 (2016)

'Gaia Data Release 2. Summary of the contents and survey properties', Gaia Collaboration, *Astron. & Astroph*, 616, A1 (2018)

'1777 variables in the Magellanic Clouds', H. S. Leavitt, *Annals of Harvard College Observatory*, 60, 87 (1908)

'Periods of 25 Variable Stars in the Small Magellanic Cloud', H. S. Leavitt, *Harvard College Observatory Circular*, 173, 1 (1912)

'Globular Clusters and the Structure of the Galactic System', H. Shapley, *Publ. Astron. Soc. Pac.*, 30, 173 (1919)

'NGC 6822, a remote stellar system', E. Hubble, *Astroph. Jour.*, 62, 409 (1925)

'Extragalactic Nebulae', E. Hubble, *Astroph. Jour.*, 64, 321 (1926)

'The Laniakea supercluster of galaxies', R. B. Tully, H. Courtois, Y. Hoffman & D. Pomerade, *Nature*, 513, 71 (2014)

CHAPTER 2

'Spectra of bright southern stars', A. J. Cannon, *Annals of Harvard College Observatory*, 28, 129 (1901)

'On the Relation Between Brightness and Spectral Type in the Pleiades', H. Rosenberg, *Astronomische Nachrichten*, 186, 71 (1910)

'Relations Between the Spectra and Other Characteristics of the Stars', H. N. Russell, *Popular Astronomy*, 22, 275 (1914)

'Stellar Atmospheres; a Contribution to the Observational Study of High Temperature in the Reversing Layers of Stars', C. Payne-Gaposchkin, Doctoral thesis, Radcliffe College (1925)

'The Internal Constitution of the Stars', A. Eddington, *The Observatory*, 43, 341 (1920)

'Energy Production in Stars', H. Bethe, *Phys. Rev.*, 55, 434 (1939)

'The Maximum Mass of Ideal White Dwarfs', S. Chandrasekhar, *Astroph. Jour.*, 74, 81 (1931)

'An extremely luminous X-ray outburst at the birth of a supernova', A. Soderberg et al., *Nature*, 453, 469 (2008)

'Cosmic rays from super-novae', W. Baade & F. Zwicky, *Proc. Natl. Acad. Sci.*, 20, 259 (1934)

'On Super-novae', W. Baade & F. Zwicky, *Proc. Natl. Acad. Sci.*, 20, 254 (1934)

'Energy Emission from a Neutron Star', F. Pacini, *Nature*, 216, 567 (1967)

'Observation of a Rapidly Pulsating Radio Source', A. Hewish, J. Bell, J. Pilkington, P. Scott & R. Collins, *Nature*, 217, 709 (1968).

'Die Feldgleichungen der Gravitation (The Field Equations of Gravitation)', A. Einstein, *Sitzungsberichte der Preussischen Akademie der Wissenschaften zu Berlin*, 844 (1915)

'Observation of Gravitational Waves from a Binary Black Hole Merger', LIGO and Virgo Collaborations, *Phys. Rev. Lett.*, 116, 061102 (2016)

'Discovery of a pulsar in a binary system', R. Hulse & J. Taylor, *Astroph. Jour.*, 195, L51 (1975)

'Multi-messenger Observations of a Binary Neutron Star Merger', B. Abbott et al., *Astroph. Jour. Lett.*, 848, L12 (2017)

'A planetary system around the millisecond pulsar PSR1257+12', A. Wolszczan & D. Frail, *Nature*, 355, 145 (1992)

'A Jupiter-mass companion to a solar-type star', M. Mayor & D. Queloz, *Nature*, 378, 355 (1995)

'Temperate Earth-sized planets transiting a nearby ultracool dwarf star', M. Gillon et al., *Nature*, 533, 221 (2016)

CHAPTER 3 ─────────────────────────────────────

'Die Rotverschiebung von extragalaktischen Nebeln (The redshift of extragalactic nebulae)', F. Zwicky, Helvetica Physica Acta, 6, 110 (1933) [Republished in English translation in *Gen. Rel. Gravit.*, 41, 207 (2009)]

'Extended rotation curves of high-luminosity spiral galaxies', V. Rubin, K. Ford & N. Thonnard, *Astroph. Jour. Lett.* 225, L107 (1978)

'The size and mass of galaxies, and the mass of the universe', P. J. Peebles, J. Ostriker, A. Yahil, *Astroph. Jour.*, 193, L1 (1974)

'Survey of galaxy redshifts. II – The large scale space distribution', M. Davis, J. Huchra, D. Latham & J. Tonry, *Astroph. Jour.*, 253, 423 (1981)

'The evolution of large-scale structure in a universe dominated by cold dark matter', M. Davis, G. Efstathiou, C. Frenk & S. White, *Astroph. Jour.*, 292, 371 (1985)

'First results from the IllustrisTNG simulations: matter and galaxy clustering', V. Springel et al., *Mon. Not. Roy. Astron. Soc.*, 475, 676 (2018)

'A Determination of the Deflection of Light by the Sun's Gravitational Field, from Observations Made at the Total Eclipse of May 29', F. Dyson, A. Eddington & C. Davidson, *Phil. Tran. Roy. Soc.*, 220, 291 (1920)

'Lens-Like Action of a Star by the Deviation of Light in the Gravitational Field', A. Einstein, *Science*, 84, 506 (1936)

'On the Masses of Nebulae and of Clusters of Nebulae', F. Zwicky, *Astroph. Jour.*, 86, 217 (1937)

'0957 + 561 A, B – Twin quasistellar objects or gravitational lens', D. Walsh, R. Carswell & R. Weymann, *Nature*, 279, 381 (1979)

'Multiple images of a highly magnified supernova formed by an early-type cluster galaxy lens', P. Kelly, *Science*, 347, 1123 (2015).

'Detection of the Free Neutrino: a Confirmation', C. Cowan, F. Reines, F. Harrison, H. Kruse & A. McGuire, *Science*, 124, 103 (1956)

'Solar Neutrinos: A Scientific Puzzle', J. Bahcall & R. Davis, *Science*, 191, 264 (1976)

'Evidence for Oscillation of Atmospheric Neutrinos', Super-Kamiokande Collaboration, *Phys. Rev. Lett.*, 81, 1562 (1998)

'Direct Evidence for Neutrino Flavor Transformation from Neutral-Current Interactions in the Sudbury Neutrino Observatory', SNO Collaboration, *Phys. Rev. Lett.*, 89, 011301 (2002)

'A Direct Empirical Proof of the Existence of Dark Matter',
   D. Clowe et al., *Astroph. Jour.*, 648, L109 (2006)

CHAPTER 4 ─────────────────────────────

'Über die Krümmung des Raumes (On the curvature of space)',
   A. Friedmann, *Zeitschrift für Physik*, 10, 377 (1922)
'Un Univers homogène de masse constante et de rayon croissant
   rendant compte de la vitesse radiale des nébuleuses extra-
   galactiques (A homogeneous universe of constant mass and
   increasing radius accounting for the radial velocity of extra-
   galactic nebulae)', G. Lemaître, *Annales de la Société Scientifique
   de Bruxelles*, A47, 49 (1927) [Partial translation in *Mon. Not.
   Roy. Astron. Soc.*, 91, 483-490 (1931)]
'Spectrographic Observations of Nebulae', V. Slipher, *Popular
   Astronomy*, 23, 21 (1915)
'A Relation between Distance and Radial Velocity among Extra-
   Galactic Nebulae', E. Hubble, *Proc. Natl. Acad. Sci.*, 15, 168 (1929)
'The extragalactic distance scale. VII – The velocity-distance
   relations in different directions and the Hubble ratio within
   and without the local supercluster', G. de Vaucouleurs & G.
   Bollinger, *Astroph. Jour.*, 233, 433 (1979)
'Steps toward the Hubble constant. VIII – The global value',
   A. Sandage & G. Tammann, *Astroph. Jour.*, 256, 339 (1982)
'Final Results from the Hubble Space Telescope Key Project to
   Measure the Hubble Constant', W. Freedman et al., *Astroph.
   Jour.*, 553, 47 (2001)
'Evolution of the Universe', R. Alpher & R. Herman, *Nature*, 162,
   774 (1948)
'A Measurement of Excess Antenna Temperature at 4080 Mc/s',
   A. Penzias & R. Wilson, *Astroph. Jour.*, 142, 419 (1965)
'Cosmic Black-Body Radiation', R. Dicke, P. J. Peebles, P. Roll &
   D. Wilkinson, *Astroph. Jour.*, 142, 414 (1965)
'Inflationary universe: A possible solution to the horizon and
   flatness problems', A. Guth, *Phys. Rev. D*, 23, 347 (1981)

'Bouncing cosmology made simple', A. Ijjas & P. Steinhardt, *Class. Quantum Grav.*, 35, 135004 (2018)

'A flat Universe from high-resolution maps of the cosmic microwave background radiation', F. de Bernardis et al., *Nature*, 404, 955 (2000)

'MAXIMA-1: A Measurement of the Cosmic Microwave Background Anisotropy on Angular Scales of 10'-5°', S. Hanany et al. *Astroph. Jour.*, 545, L5 (2000)

CHAPTER 5

'The Origin of Chemical Elements', R. A. Alpher, H. Bethe & G. Gamow, *Phys. Rev.* 73, 803 (1948)

'Primeval Helium Abundance and the Primeval Fireball', P. J. Peebles, *Phys. Rev. Lett.*, 16, 410 (1966)

'Cosmic Black-Body Radiation and Galaxy Formation', J. Silk, *Astroph. Jour.*, 151, 459 (1968)

Primeval Adiabatic Perturbation in an Expanding Universe, P. J. E. Peebles & J. Yu, *Astroph. Jour.*, 162, 815 (1970)

'Structure in the COBE differential microwave radiometer first-year maps', G. Smoot, C. Bennett, A. Kogut, E. Wright et al., *Astroph. Jour.*, 396, L1 (1992)

'Massive Black Holes as Population III Remnants', P. Madau & M. Rees, *Astroph. Jour.*, 551, L27 (2001)

'On the Density of Neutral Hydrogen in Intergalactic Space', J. Gunn & B. Peterson, *Astroph. Jour.*, 142, 1633 (1965)

'Evidence for Reionization at z ~ 6: Detection of a Gunn-Peterson Trough in a z=6.28 Quasar', R. Becker et al., *Astron. Jour.*, 122, 2850 (2001)

'Galaxy correlations on large scales', S. Maddox, G. Efstathiou, W. Sutherland & J. Loveday, *Mon. Not. Roy. Astron. Soc*, 242, 43 (1990)

'Observational Evidence from Supernovae for an Accelerating Universe and a Cosmological Constant', A. Riess et al., *Astroph. Jour.*, 116, 1009 (1998)

'Measurements of Ω and Λ from 42 High-Redshift Supernovae',
S. Perlmutter et al., *Astroph. Jour.*, 517, 565 (1999)

'The cosmological constant and cold dark matter', G. Efstathiou,
W. Sutherland & S. Maddox, *Nature*, 348, 705, 1990

'First-Year Wilkinson Microwave Anisotropy Probe (WMAP)
Observations: Determination of Cosmological Parameters',
D. Spergel et al., *Astroph. Jour. Supp.*, 148, 175 (2003)

Index